THE CAMPAIGN AND BATTLE OF MANZIKERT, 1071

by

GEORGIOS THEOTOKIS

ARC HUMANITIES PRESS

British Library Cataloguing in Publication Data
A catalogue record for this book is available from the British Library.

ISBN (hardback): 9781641894357
e-ISBN (PDF): 9781802701722

www.arc-humanities.org

Printed and bound in the UK (by CPI Group [UK] Ltd), USA (by Bookmasters), and elsewhere using print-on-demand technology.

To little Nikolas:
pick your battles wisely

Κι ο τρίτος ο απαράβαλτος, ολάκριβος της Φήμης,
ο ατρόμητος κυβερνητής, ο μέγας Καππαδόκης,
του Χαλεπιού ο πολέμαρχος, των Άδανων ο κύρης.
Του καταρράχτη του Τογρούλ τού στήθηκε χαράκι,
και κράτησε τον Αρπασλάν, των Τούρκων το σουλτάνο
που καβαλάρης άνεμος απ' τα βουνά του Πόντου
κι από τ' αρμένικα στενά χυνόταν ώς τους κάμπους
της Αντιόχειας· κι ύστερα νά! σκλάβος του σουλτάνου,
κι ύστερα πιο λυπητερός ραγιάς, πανάθλιος δούλος
μέσα στη χώρα που όριζε· κι έφαε την καταφρόνια,
του μουλαριού το λάχτισμα, το φτύσιμο του Εβραίου.
Κωστή Παλαμά, *Η Φλογέρα του Βασιλιά, Λόγος Δεύτερος*

Tarihte şanlar, şöhretler kazanmiş pek çok insanlar millî noktadan fazilete sahip değildir. Meselâ hakikaten askerî kudret sahibi olan, Moskova'ya kadar giden, yanginlar harabeler üstünden Fransiz ordusunu sürükleyip eriten Napolyon'u düşünürüz. Onun hareketleri Fransiz milletinin hakiki ve millî menfaatlerine değil, kendi cihangirane emellerini tatmin içindi. Bunu tatmin için Fransa'nin milyonlarca seçkin evlâdini eritti ve nihayet hepinizin bildiğiniz âkibete uğradi. Bizim Osmanli tarihindeki en büyük ve şanli görülen hareketleri de ayni noktadan tetkik, ayni mahiyette mukayese etmek mümkündür. (1923)

Atatürk'ün S. II, S. 161–162

CONTENTS

LIST OF ILLUSTRATIONS

Maps

Figures

INTRODUCTION

IT HAS BEEN described as a dreadful day, a most shattering and shameful defeat, a debacle, an unmitigated catastrophe, the breaking of Byzantium, the beginning of the end of medieval Hellenism, and the day that opened the gates of Anatolia to the Turks, while being pondered over whether or not it was "un désastre militaire" for the Byzantine army of the late 1060s. Whatever the case, the battle that was fought on August 26, 1071, between the forces of the Byzantine emperor Romanos IV Diogenes and the Seljuk sultan Alp Arslan is—rightly so—regarded as one of the most significant turning points in medieval history. In recent years, an increasing number of scholars specializing in the period have lent their voices in support of the view that the Battle of Manzikert was not the real military disaster as it was once portrayed, framing the upcoming collapse of the Byzantine rule in Anatolia that followed as the direct outcome of two factors: Byzantine aristocratic infighting and civil war, and the increased involvement of the Seljuks in these civil wars that "eased" their penetration into Anatolia.

Indeed, it is also my firm conviction that Romanos Diogenes's "Manzikert campaign" of 1071 may have been a strategic failure for the emperor, but it was not a tactical disaster. It should have only confirmed Turkish domination of the Armenian highlands in eastern Asia Minor, rather than almost all of Anatolia, which itself was overrun by the Turks within eight years. Therefore, even though no great military figure is said to have died in the battle, and it seems clear that the actual losses incurred were limited to the emperor's immediate retinue, the *decisive* outcome of the Battle of Manzikert ushered in an element of chaos in the geopolitical history of the Byzantine Empire in the 1070s: the civil wars of the early 1070s that followed the battle marked the political and military collapse of the empire in Anatolia, the Balkans, and Italy as well as central and eastern Asia Minor.

In the introductory part of this book found in the first three chapters, I examine the Christian and Muslim narrative sources for the battle strictly from a military perspective, and I reach several conclusions regarding the value of these sources for the history of warfare in Anatolia and the Eastern Mediterranean in the eleventh century. In Chapter 1, I focus my analysis on the chroniclers' social, religious, and educational backgrounds, the dates and places of the compilation of their works, their own sources and the information they from them, and their biases and sympathies, which all shed light on their level of impartiality as historians. This chapter will be followed by a comparative analysis of the sources strictly from a military perspective, reaching significant conclusions regarding their value as "military historians."

The major questions raised in Chapter 2 are: how accurate and detailed are their descriptions of pitched battles and clashes between enemy units during the Byzantine campaigns of 1068 to 1071 in the East? How far do contemporary narratives permit an accurate reconstruction of the chain of events, especially regarding the battlefield manoeuvres of armies in action? The reader of this section will also find an adequate analysis of the "battle-piece," along with the reasons and challenges involved in writing about warfare in ancient and medieval historiography, including a thorough examination of the "requirements" Byzantine classicizing historians had to meet for their works to be regarded as authoritative, and which ensured their survival.

Other questions asked in Chapter 3 are: to what extent are the figures our sources provide for army sizes reliable, both in absolute numbers and in the ratios given between cavalry and infantry? What is our chroniclers' knowledge of the local geography where the military operations took place, and to what extent—if at all—were they familiar with the terrain of the battles or sieges, or the campaign routes of armies which they describe? How far do their narratives permit the accurate reconstruction of a chain of events, especially about the battlefield manoeuvres of the armies in action?

Chapter 4 describes the geopolitical background of the Turkish infiltration of eastern Anatolia and Upper Mesopotamia, demonstrating that the Turkish campaigns that penetrated the region between the mid-1030s and the mid-1050s did not culminate in any major conquests for the Seljuks, nor did they undermine the imperial defences in Vaspourakan or any neighbouring areas. Rather, they undermined the Byzantine "buffer zone" of the Muslim vassal principalities in Upper Mesopotamia and northern Syria and, second, helped to propel the Seljuk state towards a supra-regional centralizing power and become a serious challenger to both Byzantium and the Fatimids of Egypt. I explain the strategic importance of the fortress-cities along the northern shore of Lake Van, and how the defensive system of the region of Vaspourakan that was based on the strategic assumption of a threat "from the south" was compromised by Turkish penetrations "from the east." Finally, I outline the many reasons that have been adduced for the Byzantine failure to stem the tide of invasions in different operational theatres after the death of Emperor Basil II in 1025, putting particular emphasis on the constant civil wars between emperors and pretenders concerning the imperial throne that led to critical recalls of eastern forces to suppress seditious movements. Such circumstances greatly weakened imperial defences in the eastern operational theatre, especially Vaspourakan, which eventually became the epicentre of the Seljuk invasion into central Anatolia.

In Chapter 5, I show how the emperor failed in both his campaigns in 1068 and in 1069 to inflict a decisive defeat on a Seljuk army, while the "central" border region between Theodosiopolis and Melitene remained wide open to Turkish raiding parties. Romanos fell short of winning control over the strategic invasion routes of the Arsanias River valley and the Upper Euphrates region around Melitene, thus allowing for an increased number of enemies to roam free between Vaasparakania, Taron, and the Diyar Bakr province. Furthermore, I connect what was happening in Vaasparakania and Upper Mesopotamia with Alp Arslan's focus in Syria because it was apparent to the Byzantine side by the summer of 1069 that the entire region of the emirate of Aleppo faced various threats from the Seljuk armies. I highlight that Alp Arslan's campaign in Syria in the winter of 1070/71 should be considered a strategic success because it significantly boosted his influence over the regions of northern Syria and the Upper Euphrates while putting further pressure on the remaining imperial outposts of Edessa and Antioch. The final section of this chapter deals with the strategic target for Diogenes's 1071 campaign, an estimate of the numbers and composition of his armies, the course of his army from

Constantinople to Theodosiopolis and Manzikert, and the critical aspect of strategic intelligence that eventually failed him.

The sixth chapter of this book deals explicitly with the tactical aspects of the Battle of Manzikert. I explain Diogenes's strategic miscalculations that are attributed to the strategic intelligence failure I explain in Chapter 5. These miscalculations resulted in the division of his army and the dispatch of a large section to Khliat while the emperor was under the impression that Alp Arslan had adopted a defensive strategy and, thus, would not march against him in Vaspourakan. From here, I disentangle the confusing events of first contact between the two armies, from the siege of Khliat by Trachaneiotes's detachments to Bryennios's and Basilakes's melees with well-disciplined and battle-hardened Seljuk mounted troops. The following sections deal with the events of Wednesday (August 24) and Thursday (August 25) before the main battle— the terrifying night spent in the imperial camp while under a blockade by the Seljuk army, followed by failed negotiations for a truce, and the emperor's desperate attempt to recall his troops from Khliat that never reached them or went unheeded (we will never know!). The final part of chapter 6 is a stage-by-stage analysis of the Battle of Manzikert that includes a presentation of the potential locations for the battlefield, the advantages and disadvantages of these locations depending on whether an army was attacking or on the defensive, and—critically—the challenges that would have posed for heavily armed units struggling to keep their discipline and maintain an unbroken line through their advance against an elusive enemy.

Chapter 7 introduces the aftermath of the Battle of Manzikert, with a description of the chaos that ensued in the imperial camp following the breaking of the lines of the imperial army and the pell-mell retreat of the units back to the camp, as described by our most important eye-witness, Michael Attaleiates. Then, I describe the reception of the story of the captivity of Emperor Diogenes by Alp Arslan in Muslim historiography of the following centuries, emphasizing specific episodes in their encounters with each other and the ideological significance for Byzantine-Turkish relations of the period. Finally, I look at the political challenges for the imperial government after the Battle of Manzikert, particularly the civil war of 1071–1072, the three-year war with Roussel of Bailleuil (1073–1076), and the multiple civil wars of 1077–1081.

The Secondary Bibliography

One of the most useful studies for this historical period that encompasses the Battle of Manzikert is a recent book by one of the most well-known scholars in the field. Alexander Daniel Beihammer's *Byzantium and the Emergence of Muslim-Turkish Anatolia, ca. 1040–1130* (London: Routledge, 2017) is a meticulous, lucid, and well-structured book that tackles one of the most complicated topics in the political history of the medieval Near East: the political evolution of Anatolia and Northern Syria during the time of the Turkish conquests. Beihammer's study is of particular importance for the period leading up to the Battle of Manzikert, as he scrupulously untangles the complex politics of this era—a challenge in its own right—and seeks to recreate the political development of the Anatolian/Northern Syrian world during the years

1040–1130.[1] Critically, Beihammer adds to the current consensus that Manzikert was not a military disaster but a catalyst that exposed the moral failure and weakness of the Byzantine ruling class to defend the empire, leading to the Seljuk Sultanate's deeper involvement in Byzantine imperial affairs, thus preparing the ground for more powerful interventions in the interior of Asia Minor.

A valuable analysis of the Battle of Manzikert and its background, which is suitable both for the general audience and for readers who are already well-versed in the period in question, is *Manzikert 1071: The Breaking of Byzantium* (Oxford: Osprey, 2013). Written by David Nicolle, a highly-respected scholar in the field of military archaeology of the eastern Mediterranean who has produced many ground-breaking books and articles on medieval and Islamic warfare, *Manzikert 1071* offers a competent examination of the historical background of the battle, the opposing commanders and their plans, and a skilful stage-by-stage analysis of the prelude and the events on the day of the battle.[2] Nicolle's section about the battle is the most detailed one (in English) after the publication of *The Dreadful Day: The Battle of Manzikert 1071* by Alfred Friendly (London: Hutchinson, 1981). Nonetheless, Friendly is not successful in his effort to present a compact survey of the internal and external Byzantine, Islamic, and more especially Turkic history. Furthermore, specialists will object to oversimplifications and tenuous generalizations, while also spotting glaring factual, geographic, and bibliographic errors.[3] Another book that deals with the military history of the period between Justinian and Diogenes is *The Road to Manzikert: Byzantine and Islamic Warfare, 527–1071* by Joshua B. Allfree, John Cairns, and Brian Todd Carey (Barnsley: Pen & Sword, 2012). The lack of variety and haphazard use of primary and secondary sources, the superficial reading of the authors about the period in question, and the subsequent shallow analysis of the battle and its aftermath make this monograph another lost opportunity for a comprehensive military history of the period leading to the Battle of Manzikert.[4]

Of direct interest to the topic of this book are two new publications that were apparently influenced by the approaching commemorations for the 950th anniversary of Alp Arslan's victory. Muharrem Kesik's new study on the Battle of Manzikert and its aftermath, titled *1071 Malazgirt Zafere Giden Yol* (Timaş Yayinlari: Istanbul, 2014), acknowledges and maintains the well-established views for this battle in both Turkish and European scientific discourse as a decisive event in the history of Anatolia and a landmark

---

1 Nicholas Morton, review of *Byzantium and the Emergence of Muslim-Turkish Anatolia, ca. 1040–1130*, by Alexander D. Beihammer, *Journal of Transcultural Medieval Studies* 5 (2018): 167–69.

2 Because they are not limited to academic audiences, the great advantage of this Osprey publication is the superb quality of the many 3-D maps and photos in each section of the book. On the downside, the restricted length of the book does not permit a thorough presentation of the opposing armies regarding their trends, their attitude, and their culture.

3 Andrew S. Ehrenkreutz, review of *The Dreadful Day: The Battle of Manzikert 1071* by Alfred Friendly, by Alfred Friendly, *Middle East Journal* 36 (1982): 600–601; Mark Wojcik, review of *The Dreadful Day: The Battle of Manzikert*, by Alfred Friendly, *UCLA Historical Journal* 3 (1982): 90–91.

4 Lucas McMahon, review of *The Road to Manzikert: Byzantine and Islamic Warfare, 527–1071*, by Brian Todd Carey, Joshua B. Allfree, and John Cairns, *The Mediaeval Journal* 4, no. 2 (2014): 4–6.

in the transition from Byzantine to Seljuk rule. Unfortunately, I was unable to consult a collective edition for both the wider and specialized audiences that came out on time for the commemoration of the Battle of Manzikert in 2021. Titled *950: Yilinda Malazgirt Zaferi ve Sultan Alp Arslan* [950 Years: Manzikert Victory and Sultan Alp Arslan] and edited by Hakan Sari and Yusuf Koşar (Istanbul: Ihlamur Kitap, 2021), this book analyzes the Battle of Manzikert and its aftermath, and includeds a discussion of topics such as the different states that warred to dominate Anatolia, the commanders for both sides, the development of the war itself and its impact on the crusades, the economy and logistics of the war, and what my Turkish colleagues call "the Turkification process of Anatolia." Less detailed about the prelude and the Battle of Manzikert, but a still useful source for the Seljuk invasions and settlement in Anatolia, are two older monographs, th first by Osman Turan (1914–1978), titled *Selçuklular Tarihi ve Türk-İslâm Medeniyeti* [History of Seljuks and Turkish-Islamic Civilization] (Istanbul: Ötüken Neşriyat, 1965/2008), and the second by Ali Sevim (1928–2013), titled *Anadolu'nun Fethi Selçuklular Dönemi* [Conquest of Anatolia: The Seljukid Period] (Ankara: Türk Tarih Kurumu Basimevi, 1988). Both authors place Alp Arslan's victory and the Seljuk penetration of Anatolia within the traditional twentieth-century Turkish school of historical thought that is deeply influenced by a key concept of Turkish nationalism, presenting Anatolia as the Turks' natural homeland (*yurdu*) and final destination after a centuries-long process of migration.[5]

Of invaluable use for the historians and archaeologists of the period is a recent edition of the Turkish Magazine *Aktüel Arkeoloji Dergisi* [Actual Archaeology Magazine] published in 2021.[6] Through the online source, the reader can access Turkish-language articles by respected Turkish academics concerning recent research and findings as to the strategy and tactics of the battle (Adnan Çevik, "Türklere Anadolu'nun Kapilarini Açan Savaş; Malazgirt" [The War that Opened the Gates of Anatolia to the Turks: Manzikert]), recent excavations in the region of Manzikert (A. Oğuzhan Karaçetin, "Malazgirt'te Savaş Alani Arkeolojisi" [Battlefield Archaeology at Malazgirt]; İlayda Ürün, Muhammed Dolmuşand Ali Metin Büyükkarakaya, "Malazgirt Ziyaret Tepe Kazilari" [Manzikert Ziyaret Tepe Excavations]), a re-evaluation of the medieval fortifications of Manzikert (Kemalettin Köroğluand Mehmet Sait Sütcü, "Malazgirt: Bir Ortaçağ Garnizonu ve Surlari" [Malazgirt: A Medieval Garrison and Walls]), and a comparison of the two rulers, Diogenes and Alp Arslan (Mustafa Alican, "Bir Savaş İki Hükümdar" [One War, Two Monarchs]).

The output in Greek about the Battle of Manzikert is far less notable both in terms of quality and quantity. The voluminous *Μάντζικερτ, Η αρχή του τέλους του μεσαιωνικού ελληνισμού* [Mantzikert: The Beginning of the End of Medieval Hellenism] (Athens: Govostis, 1996) was not penned by a historian but a judge in Greece's highest constitutional court, Nikos Tsaggas, and is a popularizing monograph with a lively narrative, but the research is based on older studies while ignoring most existing primary and secondary sources. The shorter and more academically minded *Η μάχη του Μαντζικέρτ και*

5 Beihammer, *Byzantium and the Emergence of Muslim-Turkish Anatolia*, 14.

6 *Aktüel Arkeoloji Dergisi* [Actual Archaeology Magazine] 83 (2021), https://aktuelarkeoloji.com.tr/, accessed July 29, 2022.

οι συνέπειές της στη βυζαντινή αυτοκρατορία [The Battle of Manzikert and its Consequences for the Byzantine Empire] by Sotirios P. Lekkas (Athens: Kaktos, 2018) offers nothing new to the discussion about the battle, nor does the special issue of the popularizing magazine *Στρατιωτική Ιστορία* [Military History] on *Μάντζικερτ 1071: Η αρχή του τέλους της βυζαντινής αυτοκρατορία* [Manzikert 1071: The Beginning of the End of the Byzantine Empire] (Athens, 2009), penned by Effie Ragia.

Highly valuable for the study of the period under discussion is the voluminous doctoral thesis (in Greek) of Georgios A. Leveniotes, submitted under the title *Η πολιτική κατάρρευση του Βυζαντίου στην Ανατολή: το ανατολικό σύνορο και η κεντρική Μικρά Ασία κατά το β' ήμισυ του 11ου αι* [The Political Collapse of Byzantium in the East: The Eastern Frontier and Central Asia Minor in the Second Half of the Eleventh Century] (Aristotle University of Thessalonike, 2007). Focusing largely on the administrative and military organization and the collapse of the easternmost regions of the Byzantine Empire during the second half of the eleventh century, Leveniotes's study examines the Byzantine expansion in the Armenian and Syriac regions, the organization of these regions, the settlement of compact Armenian populations in central Asia Minor, and its consequences. It also analyzes in detail the Turkish attacks on the border regions, the subsequent Turkish incursions into the interior of Asia Minor, the Battle of Manzikert and the chaotic period that followed, the strengthening of the Armenian factor, and the progressive disintegration of the administrative structures of the empire in its eastern regions. What makes Leveniotes's study so invaluable for researchers like myself who do not have specialized knowledge of sigillography or seals is his extensive use of existing sigillography material and other epigraphic evidence and archaeological findings that paint a comprehensive picture of all the imperial officials who served in the eastern regions of the empire in the eleventh century. Indeed, even though it was submitted in 2007, the only disadvantage I could find is that it has never been s translated into English, thus denying access to a wealth of research from scholars who do not have a working knowledge of Greek.

Modern historians are fortunate to have Carole Hillenbrand's excellent discussion on the Battle of Manzikert from the perspective of Arabic and Persian historiography. In her *Turkish Myth and Muslim Symbol: The Battle of Manzikert* (Edinburgh: Edinburgh University Press, 2007), the author includes a discussion concerning twelfth-century sources and continues to examine the battle through the lens of the passing centuries by including an analysis of thirteenth-, fourteenth-, and fifteenth-century material. Hillenbrand's main purpose is to demonstrate how Muslim writers "gradually came to use Manzikert as a vehicle for portraying spiritual truths and for demonstrating the inherent superiority of Islam over Christianity."[7] In addition, the translation of the sources and the commentary that supports it in the first part of the book highlights numerous "trends" in Muslim historiography dealing with the Battle of Manzikert and the legends that surrounded it. In particular, the author wants us to focus our attention on the postponement of combat until a Friday; Alp Arslan's prayer-

---

7 Carole Hillenbrand, *Turkish Myth and Muslim Symbol: The Battle of Manzikert* (Edinburgh: Edinburgh University Press, 2007), 5.

ful posture before the battle; the humiliation of the emperor when he is captured by a mere slave; and the merciful treatment of the captive emperor by Alp Arslan, whose humility and simplicity is contrasted to Romanos's pride and pomp. These trends will be highlighted in the coming chapters.

Manzikert and the Concept of Decisive Battle

In this section of the introduction, I will not attempt to provide an answer to the question of whether Manzikert was a decisive battle. Rather, I will set out my understanding of the concept of decisive battles in world history that, in a sense, will form the backdrop for consideration throughout the following chapters of this study as whether or not Manzikert was indeed one of them. However, let this introduction serve as a fair warning that this study does not go into detail concerning the institutional framework of the armies in the region of the eastern Mediterranean, nor does it break new ground in the logistics of the wars of the period. Furthermore, it does not talk about the experience of the common soldier in battle, although there is a detailed discussion on what historians call "the battle-piece"; nor does it delve into the impact of war on the (border) societies of central or eastern Anatolia, Syria, or Upper Mesopotamia. There I do not focus on a detailed analysis of the organization of the war by the different "military cultures," will be frequent and detailed enough sections to provide the necessary theoretical background to the reader.

Some may raise an eyebrow as to why I have deviated from the "fashionable" narratives of the so-called "new military history"[8] school that has had a dominant influence over historical output since the 1980s. This is a school that aspires to bring the field of military history and the socio-economic analyses of Karl Marx closer together, focusing on three main contexts: (1) the political-institutional context that covers the relationship between the political and the military institutions within a state and the degree to which an army could be used as an instrument of politics; (2) the socio-economic context, an area that includes the impact of war on societies (economic productivity, logistics, recruitment, technology, and so on) and that of societies on war; and (3) the cultural context that encompasses the interaction of warrior values with the cultural values of societies in general (glorification or condemnation of warrior values through epic poems, folk songs and tales, and so forth). To make things clear, I am not disputing the significance of matters such as administration, the institutional framework for warfare, supply systems and logistics, and societies during the war; after all, there will be detailed sections for several of the aforementioned matters in key parts of this monograph to serve as the backdrop for the better understanding of the geopolitical and military events of the period under consideration. Likewise, I do not attempt to reshape the contemporary view that requires sieges, raids, skirmishes, and ambushes to dominate medieval warfare, but, as I will show later, it is the rarity of pitched battles that makes

---

8 Stephen Morillo, *What is Military History?* (Cambridge: Cambridge University Press, 2013), 4; Stephen Morillo, Jeremy Black, and Paul Lococo, *War in World History: Society, Technology, and War from Ancient Times to the Present: Volume 1, to 1500* (New York: McGraw, 2009), ix–xi.

them invaluable historical events to study. Therefore, it is my intention to reintegrate the operational, tactical, technical, and equipment aspects of the conduct of warfare, and to give the reader a broader understanding of how significant and decisive pitched battles could be on a *macro-historical* analysis, which seeks out large, long-term trends in world history.

In *The Face of Battle*, John Keegan argues emphatically that "Battle history, or campaign history, deserves a similar primacy over all other branches of military historiography...For it is not through what armies *are* but by what they *do* that the lives of nations and individuals are changed."[9] Not wishing to compare myself to one of the most brilliant military historians of the twentieth century, I also described the pitched battle as "the ultimate 'Darwinian test'" in *Twenty Battles that Shaped Medieval Europe*,[10] wanting to emphasize its cardinal importance in the shaping of geopolitical events in different cultures and periods in human history. But what is it that makes a battle *decisive*? The answer is straightforward: *impact*! From Salamis to Lechfeld, and from Hastings to the second Battle of Kosovo, a decisive battle has long-term socio-political implications that fundamentally affect the balance of power on more than just the local level. We read in *Maurice's Strategikon* that "mistakes made in ordinary affairs can generally be remedied in a short while, but errors made in war cause lasting harm,"[11] and that is because a single frenzied and violent episode that barely lasted an entire day (as opposed to a lengthy siege operation), and provided history with a clear winner, also formed the *catalyst* that introduced an element of *chaos* in history, where 'small inputs can create very large perturbations."[12]

To better understand the last sentence, we need to consider three basic points about pitched battles: first, they were rare, a characteristic that makes them invaluable for historians interested in the operational and tactical changes of armies over time. A major trend in recent medieval military historiography has been to emphasize—and rightly so—the rarity of pitched battles and the caution and reluctance of military commanders to engage in full-scale battle unless the odds weighed heavily in their favour.[13] The

---

9 John Keegan, *The Face of Battle: A Study of Agincourt, Waterloo and the Somme* (London: Pimlico, 2004), 18.

10 Georgios Theotokis, *Twenty Battles that Shaped Medieval Europe* (Ramsbury: Robert Hale, 2019), 14.

11 George T. Dennis (ed.), *Maurice's* Strategikon*: Handbook of Byzantine Military Strategy* (Philadelphia: University of Pennsylvania Press, 1984), 88.

12 Stephen Morillo, *The Battle of Hastings: Sources and Interpretations* (Woodbridge: Boydell, 1996), xix–xx. On the element of chaos in history: George A. Reisch, "Chaos, History, and, Narrative," *History and Theory* 30 (1991): 1–20.

13 R. C. Smail, *Crusading Warfare, 1097–1193* (Cambridge: Cambridge University Press, 1995), 12–17; Steve Tibble, *The Crusader Armies, 1099–1187* (New Haven: Yale University Press, 2018), 51–66; John France, *Victory in the East: A Military History of the First Crusade* (Cambridge: Cambridge University Press, 1994), 65–66; Christopher Marshall, *Warfare in the Latin East, 1192–1291* (Cambridge: Cambridge University Press, 1992), 145–47; Jim Bradbury, "Battles in England and Normandy, 1066–1154," *Anglo-Norman Studies* 6 (1983): 1–12; John Gillingham, "Richard I and the Science of War in the Middle Ages," in *Richard Coeur de Lion: Kingship, Chivalry*

majority of this work has, for the most part, focused on the study of war and battles between the rulers of the different principalities of northwestern France, England, and the Latin East. In two very influential contributions, John Gillingham has argued that the Battle of Hastings was an unusual military event in an otherwise conventional "Norman Conquest" characterized by ravaging, war of attrition, and castle-building; strategies Duke William had learned well in Normandy in the previous decades. Gillingham has also highlighted the mistaken view of Richard I of England a century later as an impetuous military leader and a poor administrator, arguing, rather, that the king's record in war proves his "Vegetian" credentials.[14] Yet, the whole concept of the rarity of pitched battles in medieval military history and the tendency of medieval commanders to avoid them was first penned in 1956 by R. C. Smail, who demonstrated the reluctance of the crusader armies in Palestine to give battle. In his chapter on "The Latin Field Army in Action," he categorized crusader military activities into three categories: (1) campaigns without battle, (2) fighting on the march, and (3) pitched battles.[15]

In 1992, Matthew Strickland stressed the same point for the operational theatre of war between the Anglo-Norman and Scottish armies on the twelfth-century Anglo-Scottish border, where the permanent military imbalance that favoured the professional Anglo-Norman knightly elite over the native Scottish armies comprised of lightly armed infantry and small bands of elite cavalry, dictated the avoidance of battle by the latter and the aggressive pursue of a decisive clash by the former. In fact, the example of the Anglo-Scottish wars of the twelfth century has added value for the study of the Byzantine–Turkish conflict leading up to Manzikert because of the key role that Anglo-Norman castles played. Though intended to protect the countryside from ravaging they eventually failed during the most immediate and fundamental expression of hostilities, and also failed to block the invasion of the English North by the mobile and elusive Scottish and Galwegian (from Galloway) tribesmen—a comparison between the Scottish and the Seljuk invasions could yield some fascinating insights into intercultural warfare in both regions.[16]

---

and War in the Twelfth Century, ed. John Gillingham (London: Hambledon, 1994), 211–26; John Gillingham, "William the Bastard at War," in *Studies in Medieval History: Presented to R. Allen Brown*, ed. Christopher Harper-Bill, Christopher J. Holdsworth, and Janet L. Nelson (Wolfeboro: Boydell, 1989), 141–58; Matthew Strickland, "Securing the North: Invasion and the Strategy of Defense in 12th-century Anglo-Scottish Warfare," in *Anglo-Norman Warfare: Studies in Late Anglo Saxon and Anglo-Norman Military Organization and Warfare*, ed. Matthew Strickland (Woodbridge: Boydell, 1992), 208–29; Sean McGlynn, "The Myths of Medieval Warfare," *History Today* 44 (1994): 28–34; Bernard S. Bachrach, "Medieval Siege Warfare: A Reconnaissance," *Journal of Military History* 58 (1994), 119–33; Bernard S. Bachrach, "Medieval Military Historiography," in *Companion to History*, ed. Michael Bentley (London: Routledge, 1997), 203–20.

14 See previous note, and: John Gillingham, "'Up with Orthodoxy!' In Defense of Vegetian Warfare," *Journal of Medieval Military History* 2 (2004), 149–58; Stephen Morillo, "Battle Seeking: The Context and Limits of Vegetian Strategy," *Journal of Medieval Military History* 1 (2003): 21–41; Clifford J. Rogers, "The Vegetian 'Science of Warfare' in the Middle Ages," *Journal of Medieval Military History* 1 (2003): 1–19.

15 Smail, *Crusading Warfare*, 138–203.

16 Strickland, "Securing the North," 179–86. For a general typology of transcultural wars: Stephen Morillo, "A General Typology of Transcultural Wars: The Early Middle Ages and Beyond," in

Yet, even though battle avoidance was neither new nor uncommon in the lands around the Eastern Mediterranean, it has been only recently that a detailed review of war in the Late Byzantine period has concluded that pitched battles were indeed much rarer than other types of conflict, like sieges and raids that were aimed at wearing down opponents rather than destroying armies.[17] In an article published in 2006 on the different categories of wars the Byzantines fought in their long history, which is aptly titled "Byzantium, the Reluctant Warrior," Warren Treadgold argues that:

> My point is not that the Byzantines defended themselves badly, because for the most part they defended themselves very well. After all, the empire lasted longer than all its enemies. It outlived the Persian Empire, two Muslim caliphates, two Bulgarian empires, and the Seljuk Turks; even the Ottoman Turks, who brought Byzantium down, founded an empire that was to have a much shorter life. The Byzantines were probably wise to avoid major pitched battles, especially because they lost most of the few they did fight, like the battle of Adrianople against the Goths, the battle of the Yarmuk during the initial Muslim expansion, and the battles of Manzikert and Myriocephalum against the Seljuk Turks.[18]

Accordingly, we can understand why battles were rare and why the Byzantines were "probably wise to avoid them," following recommendations in the military manuals of Antiquity and the Middle Ages:

> It is preferable to subdue an enemy by famine, raids and terror, than in battle where fortune ["fortuna"] tends to have more influence than bravery.[19]
>
> To try simply to overpower the enemy in the open, hand to hand and face to face, even though you might appear to win, is an enterprise which is very risky ["της τυχούσης"] and can result in serious harm. Apart from extreme emergency, it is ridiculous to try to gain victory which is too costly and brings only empty glory.[20]
>
> It is good if your enemies are harmed either by deception or raids, or by famine; and continue to harass them more and more, but do not challenge them in open war, because luck ["της τύχης"] plays as major a role as valour in battle.[21]

Transcultural Wars from the Middle Ages to the 21st Century, ed. Hans-Henning Kortüm (Berlin: de Gruyter, 2005), 29–42.

17 Christos Makrypoulias, Taxiarches Kolias, Georgios Kardaras, "An Overview of the Armed Conflicts in Late Byzantium: Theoretical Foundations and Current Research," *Byzantina Symmeikta* 31 (2021): 177–91.

18 Warren Treadgold, "Byzantium, the Reluctant Warrior," in *Noble Ideals and Bloody Realities: Warfare in the Middle Ages, 378–1492*, ed. Niall Christie and Maya Yazigi (Leiden: Brill, 2006), 209–33 (quotation at 215).

19 Vegetius, *Epitome*, bk. 3, 26.

20 Maurice's *Strategikon*, bk. 7, preface.

21 George T. Dennis (ed.), *The Taktika of Leo VI* (Washington, DC: Dumbarton Oaks, 2010), cons. 20, para. 51, 555. See also: cons. 18, para. 121, 483 ("It is very dangerous, as we have frequently

Therefore, the rarity of battles in the pre-industrial era comes as a direct result of a hugely influential factor: *chance*! Victory in a battle does not necessarily prove the social, economic, or technological superiority of a "military culture" over another. Victory may be due to other random factors that could, and have, upset the turn of events, such as an accidental arrow shot (Hastings, 1066), an unexpected rainstorm (Dorystolon, 971), a sandstorm (Mytilene, 1092), a royal horse running astray on the battlefield causing panic among the troops (Achelous, 917), or simply an underestimation of the enemy due to engrained stereotypes (Isandlwana, 1879). The element of chance and upturning of events is what Carl von Clausewitz has called *friction*:

> [T]he only concept that more or less corresponds to the factors that distinguish real war from war on paper...This tremendous friction, which cannot, as in mechanics, be reduced to a few points, is everywhere in contact with chance, and brings about effects that cannot be measured, just because they are largely due to chance.[22]

Twelve centuries before the publication of the *On War*, we also read in Maurice's *Strategikon* about the *uncertainties of war* [τα τῶν πολέμων ἄδηλον], "[so that] the general ought to be ready, even after victory, to listen to proposals of the enemy for peace on advantageous terms."[23] A few lines later, the same author reiterates his view that "the best leader [ἄριστος ἡγεμών] is one who does not willingly engage in a hazardous and highly uncertain battle."

Finally, in a period of history when a king or a sultan was at the forefront of the fighting, the death of a leader or extensive losses on the battlefield could have a *decisive* impact in the course of history and dramatically upset the balance of power between two forces for many years or even decades—or even forever.[24] To give a handful of examples to prove this point: every history student has been taught of the sudden death of the Anglo-Saxon King Harold at Hastings, who perished as a result of a "lucky" arrow shot through his eye (historically accurate or not, that is not the point I wish to make here).[25]

---

said, for anyone to run the risk of a pitched battle, even when it seems perfectly clear that our forces far outnumber the enemy. The result of fortune is unseen").

22 Carl von Clausewitz, *On War*, ed. and trans. M. Howard and P. Paret (Princeton: Princeton University Press, 1989), 119–20. The "fog of war" is often attributed to Clausewitz, but is in fact a paraphrase of what he wrote: "War is the realm of uncertainty; three quarters of the factors on which action in war is based are wrapped in a fog of greater or lesser uncertainty." *On War*, 101.

23 Maurice's *Strategikon*, 87. Repeated in: Iulius Africanus Cesti, *The Extant Fragments*, ed. Martin Wallraff, Carlo Scardino, Laura Mecella, and Christophe Guignard, trans. William Adler (Berlin: de Gruyter, 2012), 40 ("ἄδηλον γὰρ τὸ τοῦ πολέμου τέλος, καὶ παρὰ δόξαν τὰ πολλὰ χωρεῖ"); Kekaumenos, *Στρατηγικόν*, 74 ("τῆ; νίκης καί ἥττης τό ἄδηλον"); *The Taktika of Leo VI*, cons. 20, para. 119, 577.

24 Yuval Noah Harari, "The Concept of 'Decisive Battles' in World History," *Journal of World History* 18 (2007): 251–66.

25 Michael Livingston, "The Legend that Just Won"t Die: The Arrow in King Harold's Eye," in *Medieval Warfare Special: 1066–The Battle of Hastings*, ed. Peter Konieczny, 42–55 (Zutphen: Karwansaray, 2017).

Harold's untimely death proved to be the *catalyst* that tipped the scale in favour of William, Duke of Normandy, and the Normans, ultimately changing the face of English history forever. Some fifteen years later (1081) at Dyrrachium, another Norman invader—Robert "Guiscard" Hauteville—defeated the Byzantine emperor's armies in modern Albania. But even though his Norman knights had the emperor Alexios Komnenos surrounded after he fled the battlefield, the emperor managed to escape and established a rallying point at Thessaloniki. His death would have brought the empire to the brink of a renewed civil war, just like the aftermath of the Battle of Manzikert had done ten years before (1071). And even though England was as heavily fortified as Illyria or Macedonia, the key point is that the decisive nature of the victory at Hastings in 1066 and the rapid capitulation of the Anglo-Saxons saved William from having to besiege several burghs in a war of attrition that could have crippled his forces. That was not the case for Robert Guiscard.[26]

Therefore, regardless of whether battles are trustworthy or untrustworthy assessments of historical entities and movements, they are high-profile, politically charged, and pivotal events that have every potential to change the face of entire regions in a single day. Battles are the ultimate test of any military organization's effectiveness and arguably its primary function. To repeat a sentence I quoted earlier, they are the *catalyst* that introduces an element of chaos in history, where small inputs can create very large perturbations. And that is exactly what makes them captivating and so significant—at least in my eyes!

26 Georgios Theotokis, *The Norman Campaigns in the Balkans, 1081–1108* (Woodbridge: Boydell, 2014), 215–22.

Chapter 1

CHRISTIAN AND MUSLIM SOURCES FOR THE BATTLE OF MANZIKERT

MAKING SENSE OF THE PROFESSIONAL AND CULTURAL MILIEU

HAVING A WIDE variety of primary sources at a historian's disposal can be a blessing or a curse, depending on how ready one is to peel away the multiple layers of "literary goals" and "'political agendas" to reach "true history" (as opposed to Anthony Kaldellis's "unhistory"). A historian writing about the Battle of Manzikert, therefore, must tackle a dizzying array of sources not only from both the Chalcedonian and the non-Chalcedonian historiographic traditions but also from the medieval Arabic and Persian ones. Some of them were contemporaries to the events they describe, which means they had access to stories and evidence from the recent past, while others repeated well-established historical episodes for didactic and/or propagandistic purposes several centuries after the battle; some of them were (high-ranking) civil servants, while others were religious scholars and famous preachers.

To answer the question of how we are to write the history of the Battle of Manzikert in light of what we know about our sources, we should appreciate their accounts not as collections of "facts" that can be understood as being either true or false, but rather as "literary artefacts" with their own aims, skills and aesthetic."[1] Or, as Stamatina McGrath aptly put it, "the act of composing history should be understood as a deliberate, political act with the intention of influencing how the past would inform the present and future concerns of the authors, their patrons, and their worlds."[2] Therefore, to be able to read and better evaluate the battle-narrative of another author, we need to be able to understand the frame of reference with which the author constructed narratives. Accordingly, this chapter will refer to some basic points in our authors' social and educational backgrounds, the date and place of the compilation of their works, their own sources and the way they collected their information. I will try to make sense of their biases and sympathies, and how far their "socio-political agendas" shaped accounts, paying particular attention to their treatment of the personalities and reigns of Emperor Romanos and his successors, Michael VII and Nikephoros III. This is complemented by the following chapters, which critically examine the value of the Christian and Muslim sources as "military histories" for the Battle of Manzikert, and for the second half of the eleventh century in general.

---

1 Hillenbrand, *Turkish Myth*, 112; Anthony Kaldellis, "The Manufacture of History in the Later Tenth and Eleventh Centuries: Rhetorical Templates and Narrative Ontologies," in *Proceedings of the 23rd International Congress of Byzantine Studies Belgrade, 22–27 August 2016*, ed. Smilja Marjanović-Dušanić (Belgrade: The Serbian National Committee of AIEB, 2016), 295–306.

2 Stamatina McGrath, "Warfare as Literary Narrative," in *A Companion to the Byzantine Culture of War, ca. 300–1204*, ed. Yannis Stouraitis (Leiden: Brill, 2018), 167.

The Christian Sources for the Battle of Manzikert

A famous historian of modest origins who chose to write about the tumultuous period of the eleventh century was Michael Attaleiates (1020–1084/85).[3] Having received a good secondary education in philosophy, rhetoric, and the law, he was already practicing law during the reign of Michael IV (r. 1034–41).[4] After that, he was appointed as a low-ranking judge in the capital in 1042, before entering the Senate during the reign of Constantine X Doukas (r. 1059–1067). Attaleiates had earned a prominent position in the judiciary by the time of Empress-regent Eudokia, and he was asked by Romanos IV (r. 1068–1071) to accompany him as a *krités tou stratopédou* ("judge of the army")[5] on his first (1068), second (1069, which earned him the high-ranking dignity of *patrikios*), and final (1071) campaigns against the Seljuk Turks in eastern Asia Minor. Following the blinding of Romanos, Attaleiates found himself on the side of the Doukas family, with Michael VII promoting him to the dignity of *anthypatos* and, in 1075, making him a *judge of the velum*.[6] Growing increasingly frustrated with the military and financial policies of Michael VII's chief minister Nikephoritzes, Attaleiates increasingly praised the usurper Nikephoros III Botaneiates (r. 1078–1081) in his work, a tactic that eventually earned him the rank of *magister* and *proedros* in 1079.

Attaleiates's *History* has been traditionally dated to the year 1079/80,[7] and it covers the decades between 1034 and the second year of the reign of Nikephoros III, in

---

3 On Attaleiates: Dimitris Krallis, *Serving Byzantium's Emperors, The Courtly Life and Career of Michael Attaleiates* (Cham: Palgrave, 2019); Dimitris Krallis, *Michael Attaleiates and the Politics of Imperial Decline in Eleventh Century Byzantium* (Tempe: Arizona Center for Medieval and Renaissance Studies, 2012); Leonora Neville, *Guide to Byzantine Historical Writing* (Cambridge University Press: Cambridge, 2018), 150–54; Warren Treadgold, *The Middle Byzantine Historians* (New York: Palgrave, 2013), 312–29; Apostolos Karpozilos, *Βυζαντινοί Ιστορικοί και Χρονογράφοι, (11ος–12ος αι.)* (Athens: Kanake, 2009), 3:187–238; Herbert Hunger, *Βυζαντινή λογοτεχνία: η λόγια κοσμική γραμματεία των Βυζαντινών*, trans. Taxiarchis Kolias, Katerina Syneli, G. X. Makris, Ioannis Bassis (Athens: MIET, 1997), 2:201–10.

4 In this era, many historians were trained as lawyers and had careers as judges before turning to history. Angeliki Laiou, "Law, Justice, and the Byzantine Historians: Ninth to Twelfth Centuries," in *Law and Society in Byzantium: 9th–12th Centuries*, ed. Angeliki Laiou and Dieter Simon (Washington, DC: Dumbarton Oaks, 1994): 151–86.

5 Andreas Gkoutzioukostas, "Ο κριτής του στρατοπέδου και ο κριτής του φοσσάτου," *Byzantina* 26 (2006): 79–99. Gkoutzioukostas argues in 86–87 that "there is no evidence in the sources that such an office existed, as a krités tou stratopédou is not mentioned either in the Taktikon Escorial, nor in the seals, nor in the legal texts of the 11th and 12th centuries." See also: John F. Haldon, "The Krites tou Stratopedou: A New Office for a New Situation," *Travaux et mémoires* 14 (2002): 279–86.

6 Treadgold assumes that a judge of the velum was "the most senior class of judge." *Middle Byzantine Historians*, 315. On the other hand, Gkoutzioukostas thinks that "the judges of the hippodrome and the judges of the velum were, then, 'small' or inferior judges who sat as commissioned judges or assessors." Andreas Gkoutzioukostas, "'Judges of the Velum' and 'Judges of the Hippodrome' in Thessalonike (11th c.)," *Byzantina Symmeikta* 20 (2010): 69–70.

7 Modern translations include: Anthony Kaldellis and Dimitris Krallis, *The History: Michael Attaleiates* (Cambridge, MA: Harvard University Press, 2012); Inmaculada Pérez Martín, *Historia* (Madrid: Consejo Superior de Investigaciones Científicas, 2002); and Ioannes Polemis, *Μιχαήλ*

1079/80.[8] Attaleiates was a contemporary and an eyewitness (αὐτόπτης και θεατής) of the events he recounts and explains in his work, and he repeatedly draws our attention to his investigation into the causes of major events rather than relying on "hearsay and the unreliable reports of others."[9] Moreover, his position as a *krités tou stratopédou* is likely to have allowed him privileged access into the workings of the high command and, in fact, he participated in Romanos's war council more than once during the emperor's campaigns in the East between 1068 and 1071.[10] Nonetheless, Attaleiates did depend on oral and written sources about the events he did not personally witness, and he consistently refers to reports and messages as a source of information for distant events;[11] a typical example is his compiling of conflicting reports on the fate of the emperor by eyewitnesses and informants returning to the imperial camp after the Battle of Manzikert.[12] Luckily for this research, Attaleiates's narrative becomes much more detailed after the death of Constantine X in 1067, "almost" as if he is trying to convince his audience that the decade that followed Romanos's Turkish campaigns, and the disastrous reign of Michael VII, were simply a prelude to the rise on the throne of his "hero," Nikephoros III, but I will return to that shortly.

Attaleiates was critical of all the emperors whose reigns he recounted in his work (Nikephoros III was, of course, an exception) and, like Michael Psellos, he also had a mixture of good and bad qualities to report for each of them. For example, he was highly critical of Constantine X's neglect of the (thematic) armies, and of his disastrous financial policy and strategic thinking in the East (as he saw it).[13] Yet, Attaleiates reserved great respect and admiration for Romanos IV,[14] whose goal "was not love of self but solely love of his brothers and love of true religion," and as if God was indeed listening, "the prayers of the people who wished this [Romanos to take the throne] were answered."[15] However,

---

Ατταλειάτης, Ιστορία (Kanakes: Athens, 1997). On the literary composition of the *History*: Lia Raffaella Cresci, "Cadenze narrative e interpretazione critica nell'opera storica di Michele Attaliate," *Revue des études byzantines* 49 (1991): 197–218; Carla Amande, "L'Encomio di Niceforo Botaniate nella storia di Attaliate: modelli, fonti, suggestini letterarie," *Serta Historica Antiqua* 2 (1989): 265–86.

8 For the two editions of the work: Treadgold, *Middle Byzantine Historians*, 318–19.

9 Attaleiates, *History*, 10–11, 188–89, 352–53. Macrides emphasizes that Attaleiates is not trying to impress the reader with his "autopsy mania" and his constant intervention into the narrative but, rather, "it is his reading of the signs of the impending defeat which he feels bound to record." Ruth Macrides, "The Historian in the History," in *Philhellen: Studies in Honour of Robert Browning*, ed. Costas N. Constantinides, Nikolaos M. Panagiotakes, Elizabeth Jeffreys, and Athanasios D. Angelou (Venice: Istituto Ellenico di Studi Bizantini e Postbizantini di Venezia, 1996), 209–10.

10 Kyle James Sinclair, "War Writing in Middle Byzantine Historiography: Sources, Influences and Trends" (PhD diss., University of Birmingham, 2012), 87–89.

11 Sinclair, "War Writing," 89n456.

12 Sinclair, "War Writing," 81–87.

13 Attaleiates, *History*, 139–41; Karpozilos, *Βυζαντινοί Ιστορικοί και Χρονογράφοι*, 3:190–91.

14 Karpozilos indicates that Attaleiates never mentions his hero, Nikephoros III, anywhere in the narrative of Romanos's reign: *Βυζαντινοί Ιστορικοί και Χρονογράφοι*, 3:206.

15 Attaleiates, *History*, 180–81. Jonathan Shepard considered that Attaleiates "was writing to

despite of the emperor's praiseworthy attempts to revive the ailing thematic armies and his strategic determination to block the Turkish penetrations into Anatolia,[16] the historian reserved his criticisms for Romanos's strategic decisions during the Turkish campaigns in which he also took part and was consulted—his historical narrative reads like the proverbial "I told you so" with the benefit of hindsight.[17]

The historian condemns Michael's chief minister Nikephoritzes for corruption and his evil influence on the poor Michael who is portrayed as a weakling and incompetent.[18] Attaleiates also denounces Michael VII at length for blinding Romanos, and in an attempt to exalt the blind emperor, he shows his great admiration for Diogenes's acceptance of his tragic fate. In effect, we can call the *History* a "point-by-point refutation" of Psellos's pro-Doukas account of Emperor Romanos's military career and murder, with Attaleiates's narrative of the latter reading like a "prosecutorial brief using emotion to score points against Romanos's torturers."[19] Nonetheless, as Dimitris Krallis has pointed out, if we read between the lines of the *History* we can make out a more calculating and (politically) astute individual who saw early on "where the wind was blowing," and integrated himself into the Doukas administration almost immediately after Manzikert.[20] We need to understand that Attaleiates was a man struggling to stay afloat in a stormy political environment and that loyal service to an emperor was not an indication of agreement with his policies.

Attaleiates understood that the empire had suffered devastating military setbacks in the 1060–1070s, hence his "historical lament" over the *negligentia* of the military affairs of his time, which he launches at a critical part of his *History* while narrating the first Turkish campaign of Romanos in 1068.[21] We could say that his narrative reaches

---

vindicate the fallen emperor's memory." "Byzantinorussica," *Revue des études byzantines* 33 (1975): 222–23. On the contrary, Speros Vryonis, Jr. has repeatedly upheld the pre-eminence of Attaleiates as the most reliable and objective source for the reign of Romanos. Spyros Vryonis, Jr., *Η παρακμή του μεσαιωνικού ελληνισμού στη Μικρά Ασία και η διαδικασία του εξισλαμισμού 11ος–15ος αιώνας)* [The Decline of Medieval Hellenism in Asia Minor and the Process of Islamization (11th–15th Century)], trans. Katia Galatariotou (Athens: MIET, 2008), 49, 55, 80–92; "A Personal History of the History of the Battle of Manzikert," in *Byzantine Asia Minor (6th–12th cent.)*, ed. Stelios Lampakis (Athens: National Hellenic Research Foundation, 1998), 225–44; "The Eleventh Century: Was There a Crisis in the Empire? The Decline of Quality and Quantity in the Byzantine Armed Forces," in *The Empire in Crisis (?): Byzantium in the 11th Century (1025–1081)*, ed. Vassiliki N. Vlyssidou (Athens: National Hellenic Research Foundation, 2003), 17–43. See also: Antonios Vratimos-Chatzopoulos, "The Two Expeditions of the Byzantine Emperor Romanos IV Diogenes in 1068 and 1069" (PhD diss., Cardiff University, 2005), 32–40; Hillenbrand, *Turkish Myth*, 17–19; Jean-Claude Cheynet, "Manzikert: un désastre militaire?," *Byzantion* 50 (1980): 410–38.

16 Attaleiates, *History*, 188–93.

17 Attaleiates, *History*, 232–41; Treadgold, *Middle Byzantine Historians*, 321–22; Eudoxos Tsolakes, "Ο Μιχαήλ Ατταλειάτης ως κριτικός των επιχειρήσεων και της τακτικής του πολέμου," [Michael Attaleiatis as a Critic of the Operations and Tactics of War] *Βυζαντινά* 1 (1969): 187–204.

18 Attaleiates, *History*, 326–27.

19 Krallis, *Michael Attaleiates*, 80–100 (quotations in 80 and 83).

20 Krallis, *Michael Attaleiates*, 36.

21 Attaleiates, *History*, 208–11.

its climax (and his set goal) with the lengthy '"encomium" of Nikephoros III, where the reader gets the impression with the lengthy digression into Nikephoros's rebellion and the tracing of his lineage back through Nikephoros II Phokas to Constantine I, that it was God's plan all along that propelled Attaleiates's "hero" to the throne.[22] The historian prepares the ground for the rise of Nikephoros Botaneiates by reporting on the popular discontent with the incompetent Michael VII,[23] hence implying that Nikephoros was the empire's only hope in reversing the collapse of the state, which explains his triumphal and critically bloodless rise to power.[24] After all, Attaleiates finished his *History* in 1079/80, only after he was certain that the other two main contenders for the imperial throne, Bryennios ("the Elder")[25] and Basilakes, had been taken out.

Attaleiates's *History* is the literary outcome of a well-educated individual who received a round education in philosophy, rhetoric, and law, and whose studies and scholarly life in Constantinople were marked by the intellectual brilliance of the philosopher and polymath Michael Psellos, whom Attaleiates deeply admired.[26] In the *Diataxis* of Michael Attaleiates signed in 1077, the only substantial document to survive for the founding of his unassuming Constantinopolitan monastery of Christ Panoiktirmon ("the All-Merciful"), inventories appended to the document record the holdings of Attaleiates's monastic library, some seventy-nine books in total, which offer a limited, yet valuable insight into his educational background and literary interests.[27] Secular books in the monastery's catalogue, which could also have been in Attaleiates's reading list, included Josephus's *Jewish Wars*, a peripheral perspective on Roman history during times of overwhelming power and imperial expansion (hence, a far cry from Attaleiates's period), Agathias's *History*, Procopius's *Wars* and *Secret History*, the *Chronicle of Theophanes*, the *History of Leo the Deacon* and, critically, the *Strategikon* attributed to Emperor Maurice.[28] Furthermore, Pérez Martín explained the influence of classical authors in the *History*, including Homer, Herodotus, Thucydides, Xenophon, Diodorus of Sicily, Polybius, and Dionysius of Halicarnassus.[29]

The highest-ranking historian who wrote about (a brief period of) the eleventh century was Nikephoros Bryennios (d. 1136/37).[30] Writing at the request of the Byzantine

---

22 Attaleiates uses the adjective θεόμαχος ("fighting against God") to imply that whoever turned against Nikephoros, turned against God himself: *History*, 534–35. Attaleiates's apparent fondness for Botaneiates has been called into question by Krallis, *Michael Attaleiates*, 142–57.

23 Attaleiates, *History*, 470–71.

24 Attaleiates, *History*, 496–99; Karpozilos, *Βυζαντινοί Ιστορικοί και Χρονογράφοι*, 3:206–9; Hunger, *Βυζαντινή λογοτεχνία*, 2:205–7.

25 Attaleiates deliberately portrayed Nikephoros Bryennios ("the Elder"), a rival of Botaneiates, as cowardly. Cresci, "Cadenze narrative e interpretazione," 202–18. See the detailed footnote in: Sinclair, "War Writing," 90n465.

26 Krallis, *Michael Attaleiates*, 4–13.

27 Krallis, *Michael Attaleiates*, 45–52.

28 Krallis, *Michael Attaleiates*, 51.

29 Pérez Martín, *Historia*, li–liv.

30 On Nikephoros Bryennios: Leonora Neville, *Heroes and Romans in Twelfth-century Byzantium:*

empress Eirene Doukaina (1066–1123/33), wife of the emperor Alexios Komnenos (r. 1081–1118) and niece of the Caesar John Doukas, Bryennios was descended from a prominent aristocratic family from Thrace with vast estates around Adrianople, where he was probably born around 1080. He had a distinguished great-grandfather who had performed valiantly in the war against the Patzinaks in 1050 and, later, took part in the 1057 *coup d'état* that overthrew Michael VI (r. 1056–1057). Nevertheless, there is no consensus as to whether Bryennios was the son or grandson of the famous namesake and pretender to the throne, Nikephoros Bryennios ("the Elder").[31]

In 1096, Bryennios moved to the imperial palace after marrying Emperor Alexios's first-born child, the famous princess-historian Anna, which confirms Alexios's policy of connecting his family to other high-profile aristocratic houses at the time.[32] Bryennios enjoyed imperial favour and was awarded the high ranks of *panhypersebastos* (1096) and caesar (1111), while he was also trusted with different military commands against the Normans (1108), the Patzinaks (1115), the Seljuks (1116) and in many other occasions during Ioannes Komnenos's reign (r. 1118–1143). Finally, Bryennios seems to have neither supported nor tried to prevent his wife's attempt (and with Eirene Doukaina's blessing) to seize the throne from her younger brother John.

Bryennios's *Hyle Historias*[33] [Material for History][34] covers the short but tumultuous period between 1070 and 1079 and was probably compiled between 1118 (the year of Alexios's death) and shortly before the author died in 1137/38.[35] The literary sources

---

The "Material for History" of Nikephoros Bryennios (Cambridge: Cambridge University Press, 2012), 15–28; Leonora Neville, *Guide to Byzantine Historical Writing*, 169–73; Treadgold, *Middle Byzantine Historians*, 344–54; Karpozilos, *Βυζαντινοί Ιστορικοί και Χρονογράφοι*, 3:357–69; Hunger, Βυζαντινή λογοτεχνία, 2:217–19; Antonio Carile, "Il 'Cesare' Niceforo Briennio," *Aevum* 42 (1968): 429–54; Antonio Carile, "Il problema della identificazione del Cesare Niceforo Briennio," *Aevum* 38 (1964): 74–83.

31 Treadgold, *Middle Byzantine Historians*, 344; Karpozilos, Βυζαντινοί Ιστορικοί και Χρονογράφοι, 3:357–58; Hunger, Βυζαντινή λογοτεχνία, 2:217–18; Carile, "Il 'Cesare' Niceforo Briennio," 429–54; Carile, "Il problema," 74–83.

32 Neville, *Heroes and Romans*, 104–11; Peter Frankopan, "Kinship and the Distribution of Power in Komnenian Byzantium," *English Historical Review* 122 (2007): 1–34.

33 Modern translations include: María Salud Baldrich López, *Nicéforo Brienio: Materia de Historia* (Granada: Centro de Estudios Bizantinos, Neogriegos, y Chipriotas, 2012); Demetres Tsougkarakes and Despoina Tsouklidou, *Νικηφόρος Βρυέννιος, Ύλη Ιστορίας* (Athens: Kanake, 1996); Paul Gautier, *Nicephore Bryennios Histoire* (Brussels: Byzantion, 1975).

34 On the Hyle Historias: Leonora Neville, "Enemies of Empire in the Muses: Nikephoros Bryennios' Material for History, and Anna Komnene's *Alexiad*," in *Byzantine Authors and their Times*, ed. Vassiliki N. Vlyssidou (Athens: National Hellenic Research Foundation, 2021), 254–65; Neville, *Heroes and Romans in Twelfth-Century Byzantium*; Elizabeth M. Jeffreys, "Nikephoros Bryennios Reconsidered," in *Η αυτοκρατορία σε κρίση: Το Βυζάντιο τον 11ο αι (1025–1081)* [The Empire in Crisis(?): Byzantium in the Eleventh Century (1025–1081)], ed. Vassiliki N. Vlyssidou (Athens: National Hellenic Research Foundation, 2003), 201–14.

35 Bryennios may have completed his first two books during the reign of Alexios Komnenos. Karpozilos, *Βυζαντινοί Ιστορικοί και Χρονογράφοι*, 3:365; Hunger, *Βυζαντινή λογοτεχνία*, 2:220; Neville, *Heroes and Romans*, 27–28.

for the first book of the *Hyle* were Attaleiates, Skylitzes, and Psellos, while only the first two were relied on for the remaining three books of the work.[36] Judging by the intimate knowledge of Doukas's life and career that is displayed in the *Hyle*, Leonora Neville has argued that Bryennios drew on a now-lost history of the 1070s from the perspective of Caesar John Doukas.[37] Similarly, some episodes are told from the perspective of George Paleologos, which has led Neville arguing for the historian's use of another memoir-like source of the famous general.[38] Oral sources of the *Hyle* would have included stories told to him by elderly members and retainers of the Doukai, Komnenoi, and Bryennioi families, and Jonathan Shepard depicts Nikephoros Bryennios "the Elder" as "a first-hand, if biased, informant."[39] James Howard-Johnston has also put forward the intriguing theory that, besides the four books we have of the *Hyle Historias*, Nikephoros had also collected extensive materials for later books he planned to write that were eventually used by his wife Anna in her own historical masterpiece.[40]

Herbert Hunger has described the *Hyle Historias* as "a history of the Komnenoi and other illustrious families, like the Doukas and the Bryennioi, in the form of memoirs."[41] Therefore, readers can be sure that under the screen of a polite eulogy they will find only "veiled" criticism of the Komnenian and Bryennioi family policies in Nikephoros's narrative. In fact, the historian clearly differentiates his viewpoint from that of his sources whenever there is a reference to Alexios Komnenos, John and Andronikos Doukas, or his other hero, Nikephoros Bryennios ("the Elder").[42] A typical example is Bryennios's muffling of the critical role of Andronikos Doukas (Eirene Doukaina's father) and Nikephoros Bryennios "the Elder" at Manzikert, where he pins almost all the blame for the defeat on Nikephoros Basilakes.[43] Moreover, Bryennios heavily criticizes Romanos IV's rejection of "sound" strategic advice by Trachaniotes and Bryennios on the eve of the battle, in a similar "I told you so" narrative with the benefit of hindsight that we also see in Attaleiates.[44] Finally, in an excerpt that parallels Psellos's criticism of Constantine IX in the

---

36 Neville, *Heroes and Romans*, 46–59; A. Carile, "La Ὕλη Ἱστοριῶν del Cesare Niceforo Briennio," *Aevum* 43 (1969): 56–87 and 235–82.

37 Neville, *Heroes and Romans*, 49–59; Neville, "A History of the Caesar John Doukas in Nikephoros Bryennios' Material for History?," *Byzantine and Modern Greek Studies* 32 (2008): 168–88.

38 Neville, *Heroes and Romans*, 46–47.

39 Shepard, "Byzantinorussica," 221.

40 James Howard-Johnston, "Anna Komnene and the *Alexiad*," in *Alexios I Komnenos*, ed. Margaret Mullett and Dion Smythe (Belfast: Priory, 1996), 260–302. This view has been refuted by: Ruth Macrides, "The Pen and the Sword: Who Wrote the *Alexiad*?," in *Anna Komnene and Her Times*, ed. Thalia Gouma-Peterson (New York: Garland, 2000), 63–81; Jeffreys, "Nikephoros Bryennios Reconsidered," 201–14.

41 Hunger, Βυζαντινή λογοτεχνία, 3:220; Tsougkarakes, *Ύλη Ιστορίας*, 17.

42 Neville, *Heroes and Romans*, 121–38.

43 Bryennios, *Ύλη Ιστορίας*, 84–85, 87–93; Treadgold, *Middle Byzantine Historians*, 349–50; Karpozilos, *Βυζαντινοί Ιστορικοί και Χρονογράφοι*, 3:360–61; Hunger, *Βυζαντινή λογοτεχνία*, 2:220; Carile, "La Ὕλη Ἱστοριῶν," 62–66.

44 Bryennios, *Ύλη Ιστορίας*, 79–81.

Chronographia,Bryennios upholds the extent of the disaster in the 1070's that Psellos and Attaleiates downplay, which he blames on the policies of Romanos IV, on Michael VII's chief minister Nikephoritzes (but not on Michael himself), and on Nikephoros III.[45]

Bryennios was another well-educated historian for whom, however, we know far less about his reading background and literary interests than we do about Attaleiates. Individual quotations and allusions from the *Hyle Historias* suggest that Nikephoros firmly situated his work within the tradition of Greek and Roman historiography, with which he had engaged and was very well familiar, although modern scholars are left in the dark as to what works he actually read. Neville has highlighted that the relative absence of Christian literature from the *Hyle* conforms to the style of classical historiography.[46] The author often blends Homeric lines into his descriptions of battles and single-combats, while much of his military vocabulary of camping and marching is drawn from a key stylistic model, Xenophon's *Anabasis*. Other historical and literary models employed by Bryennios include Dionysius of Halicarnassus, Plutarch, Polybius, and Procopius.[47]

A prominent place among the chronographers of the twelfth century belongs to John Zonaras (ca. 1070s–second half of twelfth century).[48] Although we know surprisingly little about him, Zonaras's works provide at least some basic information about his life and career; we can be sure that he was a tweflth-century judge and bureaucrat who became a monk. Coming from a family that held high-ranking judicial and secretarial offices since the eleventh century, Zonaras's promising career prospered during the reigns of the first two Komnenoi emperors. He rose to the office of head of the imperial chancery as *protoasecretis* during the reign of Alexios I, followed by the judicial office of *droungarios tes viglas* that propelled him to the head of the civil courts in Constantinople under John II. Because he grew increasingly disaffected with the civil administration of John II's regime,[49] and/or because he was side-lined as a member of Anna Komnena's

---

45 Bryennios, *Ὕλη Ιστορίας*, 254–57; Constantine Sathas ed., *The History of Psellus* (London: Methuen, 1899), bk. 6, ln. 29, 105.

46 Neville, *Heroes and Romans*, 39.

47 Neville, *Heroes and Romans*, 40–45.

48 On Zonaras and his work: Neville, *Guide to Byzantine Historical Writing*, 191–99; Christopher Mallan, "The Historian John Zonaras: Some Observations on His Sources and Methods," in *Les historiens grecs et romains: entre sources et modèles*, ed. Olivier Devillers and Breno Sebastiani (Bordeaux: Ausonius, 2018), 353–66; Theofili Kampianaki, "John Zonaras' Epitome of Histories (12th cent.): A Compendium of Jewish–Roman History and Its Readers" (PhD diss., University of Oxford, 2017); Treadgold, *Middle Byzantine Historians*, 388–99; Karpozilos, *Βυζαντινοί Ιστορικοί και Χρονογράφοι*, 3:465–89; Thomas Banchich, *Eugene Lane, The History of Zonaras: From Alexander Severus to the Death of Theodosius the Great* (London: Routledge, 2009), 1–19; Hunger, *Βυζαντινή λογοτεχνία*, 2:246–50; Iordanes Gregoriades, *Linguistic and Literary Studies in the Epitome Historion of John Zonaras* (Thessaloniki: Centre for Byzantine Research, 1998); Gregoriades, *Ιωάννης Ζωναράς, Επιτομή Ιστοριών* (Kanake: Athens, 1995), 1:19–45.

49 Treadgold, *Middle Byzantine Historians*, 390, 392.

inner circle,[50] Zonaras eventually retired to a monastery on an island in the Sea of Marmara (either in the early 1120s or sometime in the 1130s),[51] where he wrote his work.

Although Zonaras's *Epitome Historion* begins with the Creation and ends with the death of Alexios Komnenos in 1118, it lacks many of the features commonly found in the Byzantine chronicle tradition. That is because it uses classical Greek and is secular in outlook despite the fact it was written in a monastic environment. Needless to say, we cannot be sure as to the exact period when Zonaras wrote his work, but some critical comments about Ioannes Komnenos's behaviour and policies have prompted historians to speculate that he finished his *Epitome* shortly after John died in 1143,[52] or even much later in the reign of Manuel Komnenos (r. 1143–1180).[53]

Zonaras worked with a vast array of sources, yet he was more thoughtful and critical in the way he handled his information compared to many of his past colleagues. Neville argued that "while Zonaras paraphrased his source texts more often than copying directly, he stayed close enough to the original," while Karpozilos described the outcome of Zonaras's writing as "impressive...distinguished by his objectivity, the accuracy of information...and the distance/independence he keeps from his sources."[54] Zonaras draws the majority of his information for his Books 16–18, which cover the period from the reign of Theophilos to Alexios I, from Malalas, George the monk, the lost histories of Theodore of Side and Theodore of Sebastea, Skylitzes, and Psellos.[55] Interestingly, the sole part of the *Epitome* that Zonaras produced himself during the reign of Alexios I, which includes information from well-informed eyewitnesses (officials and friends), thus re-affirming the great historical value of his work for the period of the Komnenoi that makes it comparable only to the other contemporary masterpiece of the twelfth century, Anna's *Alexiad*.

Eric McGeer, who edited the most recent translation into English of *The Continuation of the Chronicle of John Skylitzes*, made the pertinent point in his introduction on the authorship of the work that "it is the rule rather than the exception for Byzantine historical texts to have come down in more than one version as a result of interpolations, revisions, or extensions added by the author or by later hands."[56] Although both the *History*

50 Banchich, *The History of Zonaras*, 3; Gregoriades, *Επιτομή Ιστοριών*, bk. 1, 23.

51 Banchich, *The History of Zonaras*, 7; Treadgold, *Middle Byzantine Historians*, 392.

52 Treadgold, *Middle Byzantine Historians*, 389–90; Karpozilos, *Βυζαντινοί Ιστορικοί και Χρονογράφοι*, 3:467.

53 Gregoriades, *Επιτομή Ιστοριών*, bk. 22–23.

54 Neville, *Guide to Byzantine Historical Writing*, 191; Karpozilos, *Βυζαντινοί Ιστορικοί και Χρονογράφοι*, 3:486.

55 Neville, *Guide to Byzantine Historical Writing*, 196; Treadgold, *Middle Byzantine Historians*, 397; Karpozilos, *Βυζαντινοί Ιστορικοί και Χρονογράφοι*, 3:480–85.

56 Modern translations of Skylitzes's Continuation include: Eric McGeer ed., *Byzantium in the Time of Troubles: The Continuation of the Chronicle of John Skylitzes (1057–1079)* (Leiden: Brill, 2020); Eudoxos Tsolakes ed., *Σύνοψις Ἱστοριῶν (Τόμος Α')* (Athens: Kanake, 2012); John Wortley ed., *John Skylitzes: A Synopsis of Byzantine History, 811–1057* (Cambridge: Cambridge University Press, 2010); Hans Thurn ed., *Byzanz wieder ein Weltreich: das Zeitalter der makedonischen Dynastie nach*

of Michael Attaleiates and the *Chronographia* of Michael Psellos have survived in revised and expanded versions, there is still no consensus as to whether Skylitzes wrote the *Continuation*. For this research, I find myself in unison with those scholars who maintain that the *Continuation* was written by Skylitzes himself, likely at a slightly later date than the *Synopsis*.[57]

Unlike his well-known contemporaries Psellos and Attaleiates, we know surprisingly little about John Skylitzes's life and career. Skylitzes (early 1040s-first quarter twelfth century)[58] is called a "Thrakesios" in George Kedrenos's *Synopsis Historion* (compiled around 1115) and in the *Epitome Historion* of John Zonaras, which is apparently in reference to his family's origins in the Thrakesion theme. He may have been born sometime in the early 1040s and, like Attaleiates and Psellos before him, he made full use of his good secondary education in Constantinople to carve out a career in the judiciary. Skylitzes's career reached its peak during the reign of Alexios I, because by 1090 he had been awarded the high rank of *proedros* and was promoted to the prominent office of *Eparch of the City*, thus considered the supreme judge in Constantinople and its vicinity and the head of the police responsible for keeping order in the city. Skylitzes is also cited in two legal documents dated 1090 and 1092 as a *droungarios tes viglas*, a high-ranking judicial office and head of the civil courts in Constantinople, which was accompanied by the even higher rank of *kouropalates*. Skylitzes earned the more illustrious rank of *protovestiarios* only after the second edition of his *Synopsis* appeared around 1105, although this may have been just the rank of *protovestes* according to Seibt,[59] which, nonetheless, assures the high esteem and prominent position Skylitzes enjoyed in the Komnenian court for a man unrelated to the imperial family by blood. He probably had retired from the judiciary already in 1094.

---

dem Geschichtswerk des Johannes Skylitzes (Styria: Graz, 1983); Eudoxos Tsolakes ed., *Ἡ Συνέχεια τῆς Χρονογραφίας τοῦ Ἰωάννου Σκυλίτση* (Thessaloniki: Hetaireia Makedonikon Spoudon, 1968).

57 McGeer, *Continuation*, 8–10; Eudoxos Tsolakes, "Συνέχειας Συνέχεια," *Byzantina Symmeikta* 25 (2015): 115–42; Treadgold, *Middle Byzantine Historians*, 330–32; Flusin, "Re-writing History: John Skylitzes' *Synopsis Historion*," in *John Skylitzes: A Synopsis of Byzantine History, 811–1057*, xxxi; Holmes, *Basil II*, 83; Karpozilos, Βυζαντινοί Ιστορικοί και Χρονογράφοι, 3:307–11; Hunger, *Βυζαντινή λογοτεχνία*, 2:214; Werner Seibt, "Ioannes Skylitzes. Zur Person des Chronisten," *Jahrbuch der österreichischen Byzantinistik* 25 (1976): 80–85; Tsolakes, *Ἡ Συνέχεια*, 76–99. By contrast, Alexander Kazhdan, Eirene-Sophia Kiapidou, Carl de Boor, and Gyula Moravcsik believe that the Continuation was written by a different author: Neville, *Guide to Byzantine Historical Writing*, 157n15.

58 On Skylitzes: McGeer, *Continuation*, 5–7; Treadgold, *Middle Byzantine Historians*, 329–32; Eirene-Sofia Kiapidou, *Η Σύνοψη Ιστοριών του Ιωάννη Σκυλίτζη και οι πηγές της (811–1057) Συμβολή στην Βυζαντινή Ιστοριογραφία κατά τον ΙΑ΄αιώνα* (Athens: Kanake, 2010); Catherine Holmes, *Basil II and the Governance of Empire (976–1025)* (Oxford: Oxford University Press, 2005); Jean-Claude Cheynet, "Introduction: John Skylitzes, the Author and his Family," in *John Skylitzes: A Synopsis of Byzantine History, 811–1057*, ix–xi; Holmes, *Basil II*, 80–91; Karpozilos, *Βυζαντινοί Ιστορικοί και Χρονογράφοι*, 3:240–41; Hunger, *Βυζαντινή λογοτεχνία*, 2:210–11; Seibt, "Ioannes Skylitzes," 80–85; Tsolakes, *Ἡ Συνέχεια*, 76–99.

59 Seibt, "Ioannes Skylitzes," 83–84.

Around the end of his introduction to the *Synopsis* (Greek: "a seeing altogether, general view"), Skylitzes claims to be a fairly accurate summarizer (an "epitomator") of previous works rather than a historian with original views on past events. And for that reason, his professional experience in the judiciary could not have equipped him any better for what he aspired to produce. As McGeer rightly emphasized, Skylitzes's legal training and long experience in updating and condensing laws proved an invaluable experience in his "dispassionate use of his sources, his concern with the legitimacy and policies of each emperor, his attention to plots and military rebellions, his views on the legality and propriety of imperial marriages, his inclusion of opposing interpretations of events, and his awareness of the motives and the realities behind appearances";[60] in a nutshell, he felt competent as "a judge of characters" and at weighing (historical) evidence. Cheynet noted the unmistakable parallels between Skylitzes's explanation of the Byzantine defeat at Achelous, in 917, which he blamed on Romanos Lekapenos's abandonment of Leo Phokas, and that of Andronikos Doukas's betrayal of Romanos IV at Manzikert; but I will come back to that shortly.[61]

Skylitzes's *Synopsis Historion* is a comprehensive compilation of political narratives from older histories that covers the period from 811 to 1057, which he extended about a decade later to cover the period up to 1079 in what is known as *The Continuation of the Chronicle of John Skylitzes.* Both works were finished during the first half of Alexios I's reign, with the year 1095 considered the most likely *terminus ante quem* for the *Synopsis*, and the year 1105 for the *Continuation*.[62] Historians agree that the main written sources for the *Continuation* were the *History* of Attaleiates and (less) the *Chronographia* of Psellos, interspersed with oral testimonies by eyewitnesses the closer Skylitzes moves to the period of writing.[63] And the reason why Skylitzes wrote his *Continuation* was the fact that he was unaware of Attaleiates's *History* and Psellos's *Chronographia* before he published his *Synopsis*, hence his decision to make critical use of their material and add into what he had already written.[64]

60 McGeer, *Continuation*, 7–8.

61 Jean-Claude Cheynet, "Jean Skylitzès, lecteur des chroniqueurs du Xe siècle," in *Remanier, métaphraser: fonctions et techniques de la réécriture dans le monde byzantine*, ed. Smilja Marjanovic-Dušanic and Bernard Flusin (Belgrade: University of Belgrade, 2011), 117–18.

62 Treadgold, *Middle Byzantine Historians*, 331–32; Jonathan Shepard, "Memoirs as Manifesto: The Rhetoric of Katakalon Kekaumenos" in *Reading in the Byzantine Empire and Beyond*, ed. Teresa Shawcross and Ida Toth (Cambridge: Cambridge University Press, 2018), 185–214, esp. 187–89. Countering Kiapidou's arguments for a much earlier dating of the *Synopsis* in the 1060s: Kiapidou, *Η Σύνοψη Ιστοριών*, 125–36.

63 McGeer, *Continuation*, 13; Karpozilos, *Βυζαντινοί Ιστορικοί και Χρονογράφοι*, 3:313–15; Hunger, *Βυζαντινή λογοτεχνία*, 2:214; Tsolakes, *Ἡ Συνέχεια*, 61–74. An example of oral testimony of the great earthquake of 1063 that affected the area around the Sea of Marmara: "For two years thereafter earthquakes recurred from time to time, something that not even the very old could recall happening." Skylitzes, *Continuation*, 66–67.

64 Treadgold, *Middle Byzantine Historians*, 338; Karpozilos, *Βυζαντινοί Ιστορικοί και Χρονογράφοι*, 3:313–20; Hunger, Βυζαντινή *λογοτεχνία*, 3:214; Tsolakes, *Ἡ Συνέχεια*, 61–74.

Skylitzes had to tread a fine line between flattery and criticism of the Komnenoi regime. That was because he was writing the *Synopsis* during a peculiar political period, when Alexios I's son John had been crowned co-emperor only in 1092, and the emperor was still challenged by pretenders like Nikephoros Diogenes, the "porphyrogennitos" son of the emperor Romanos IV, who conspired against him in 1094 and was involved numerous confidants and relatives of Alexios, including his younger brother, Adrianos.[65] In the *Continuation*, Skylitzes shows he is very much aware of the political considerations of the Komnenoi's "matriarch" and Alexios's mother, Anna Dalassene, who undoubtedly hated the Doukai and considered the period between the death of Emperor Isaac I in 1059 and the rise to the throne of her son in 1081 as a sort of an "interregnum."[66] She even supported Eudokia and Romanos Diogenes against the Doukas family in 1067, which earned her eldest son Manuel the dignity of *kouropalates* and the office of *strategos autokrator* in the East by Emperor Romanos.[67] Yet, Skylitzes also had to consider the political sensitivities of Eirene Doukaina (1066–1123/33), wife of the emperor Alexios Komnenos (r. 1081–1118) and niece of the caesar John Doukas. Although Eirene remained in the shadow of her mother-in-law until the latter's death (probably in 1100), by the early 1100s she was apparently enjoying Alexios's confidence and had become his constant companion.[68]

Bearing all this in mind, where one stood on the matter of Romanos Diogenes relied upon one's sympathies in the larger rivalry between the Komnenoi and Doukai. Since Skylitzes would have had to count on the accounts of Attaleiates and Psellos to paint a picture of the reign of Romanos, it is easier to imagine why he relied extensively on Attaleiates in his *Continuation*; Skylitzes's account is full of great respect and admiration for the blinded emperor, who did his utmost to reverse the empire's military and economic decline but who was eventually betrayed by his generals. Skylitzes also includes the emperor's promotion of the Komnenoi and a portrayal of the emerging Alexios Komnenos in a favourable light. Conversely, the *Chronographia* would have made a much less attractive reading to a pro-Komnenian author, as it was written by a man who implies that he played an important part in replacing Isaac Komnenos with Constantine X Doukas (r. 1059–1067).[69] He also notes that he opposed restoring Romanos to the throne after he returned from his captivity. Furthermore, Psellos's flattery of Michael VII is shamelessly conspicuous in the supplement to the *Chronographia* that he composed in 1074, and Skylitzes severely criticizes him for his influence over the young emperor.[70] Yet, it should be emphasized that Skylitzes did not blindly copy Attaleiates's account of

65 Peter Frankopan, "Challenges to Imperial Authority in the Reign of Alexios I Komnemos: the Conspiracy of Nikephoros Diogenes," *Byzantinoslavica* 64 (2006): 257–74.

66 McGeer, *Continuation*, 14.

67 Lynda Garland, *Byzantine Empresses: Women and Power in Byzantium,* AD *527–1204* (London: Routledge, 1999), 187.

68 Garland, *Byzantine Empresses*, 193.

69 Psellos, *The History*, bk. 7, 91.

70 McGeer, *Continuation*, 21n72; Karpozilos, *Βυζαντινοί Ιστορικοί και Χρονογράφοι*, 3:317.

Romanos, but rather re-worked it to fit his own ends.[71] One example of this is Skylitzes's absolution of the aged emperor for his failure to block Turkish penetrations into Anatolia, instead pinning the blame on the Doukai (especially on Andronikos;) who constantly sought to undermine him.

Arguably one of the most important historiographical narratives of twelfth-century Syria, *The Chronicle of Matthew of Edessa*[72] is a fundamental source for the region of the Near East during the period of the early Crusades. The very few things we know about the author of the *Chronicle* is that he was an Armenian called Matt'ēos Uṛhayets'i (known in English as Matthew of Edessa; the third quarter of the eleventh century—shortly after 1137), a low-ranking priest and a superior of a monastery who lived for most of his life in the Syrian city of Edessa (modern Urfa), about a lifetime after the fall of the Armenian kingdom in the mid- eleventh century. Matthew's work is arranged annalistically and it covers the period starting in 952–953AD (451 of the Armenian era), and although we know that he intended to take his history up to the year 1131–1132AD (580 of the Armenian era), the last entry that can be attributed to him is the one for the year 1128–29 (577 of the Armenian era), which was probably written in late 1137; he likely died shortly thereafter. It was an otherwise unknown priest named Gregory from the neighbouring town of Kesun who extended the work until 1162–1163 (611 of the Armenian era).

Matthew explains in the preface to a second book that covers the years between 1053 and 1102 (502 to 550 of the Armenian era), that he intended to "write down the violent events of these past times, namely the horrible punishment which the Armenian nation endured at the hands of the long-haired and abominable Eghimnats'ik', the nation of the Turks, and their brothers, the Romans."[73] In spite of the unmistakable anti-Byzantine and anti-Turkish feeling in the *Chronicle*, historians have argued that these images of all non-Armenians in Matthew's work make sense only within the author's wider apocalyptic viewpoint, an essential part of Armenian historiography with which he was quite familiar.[74] These Biblical images drew largely on the eleventh-century prophesies of the *vardapet* (scholarly priest) Yovhannēs Kozeṙn, and encapsulate Matthew's philosophy of history that tries to make sense of the inexplicable (from Matthew's perspective) disasters brought by the Turkish expansion: an image of the Armenian peoples as God's "chosen" who must undergo a period of suffering so that they are cleansed of their sins; but

---

71 McGeer, *Continuation*, 19–21; Treadgold, *Middle Byzantine Historians*, 279–80; Karpozilos, *Βυζαντινοί Ιστορικοί και Χρονογράφοι*, 3:315–17.

72 For Matthew and his Chronicle: Tara L. Andrews, *Matt'eos Urhayec'i and His Chronicle: History as Apocalypse in a Crossroads of Cultures* (Leiden: Brill, 2017); Christopher MacEvitt, "The Chronicle of Matthew of Edessa: Apocalypse, the First Crusade, and the Armenian Diaspora," *Dumbarton Oaks Papers* 61 (2007): 157–81; Ara Edmond Dostourian, "The Chronicle of Matthew of Edessa (Matteos Urhayetsi)," in *Armenian Tigranakert and Diyarbakir/Edessa*, ed. R. G. Hovannisian (Costa Mesa: Mazda, 2006), 155–64; Ara Dostourian , ed., *Armenia and the Crusades: Tenth to Twelfth Centuries. The Chronicle of Matthew of Edessa* (Lanham: University of America Press, 1993).

73 Matthew of Edessa, *Chronicle*, 83.

74 Andrews, *Matt'eos Urhayec'i*, 23–43; MacEvitt, "The Chronicle of Matthew of Edessa," 167–70, 175–79.

they could look forward to eventual redemption through God's mercy, while the infidel tyrants who oppress them would be expelled.

Matthew is not only concerned about the Armenians, for his stated aim as he commences his second book is to write about "the three nations," meaning the Armenians, the Byzantines, and the Turks. Nonetheless, the overwhelming feeling of dislike and betrayal he shares for the Byzantine people and their leadership stems, according to Tara Andrews, "from the perception that, by the mid-eleventh century, the Byzantine Empire had taken on a commitment to be the "protector" of the Armenian kingdoms." It was Basil II who was viewed as the personification of that commitment, which at the time Matthew was writing had failed the Armenian nation.[75] As Matthew used his work to help frame the context of the prophetic course of events for the Armenians, his portrayal of the history of Byzantium is dominated by contrasting models of leadership: on the one hand, the warrior-emperors like Nikephoros II, John I, and Basil II, who expanded the empire and took the Syrian and Armenian Christians under their explicit protection, and on the other those who succeeded Basil II and failed miserably in the same task because they proved to be weak, corrupt, and critically intolerant of the non-Chalcedonian Armenians.

Matthew has few good things to say about Basil II's successors.[76] He is unmistakably hostile to Constantine IX Monomachos (r. 1042–1055), under whose reign Ani was incorporated into the empire, and he describes the period under Isaac Komnenos (r. 1057–1059) as one of political bickering and military defeat. He also launches a highly emotional denunciation of Byzantium's policy of "abandoning" the region between Melitene and Taron due to the Turkish incursions following the sack of Melitene (1058) and Sebasteia (1059), a policy that soon developed into a "war against the Armenians in another way; they [Romans] began to criticize their religious beliefs."[77] Yet, despite the fact that Armenians made up the bulk of Romanos IV's armies of the period, Matthew is largely unsympathetic towards Diogenes and expands upon the claims by Aristakēs Lastivertc'i that the emperor was influenced by the denunciations of certain Greeks and, despite the intervention of other Greeks alongside high Armenian dignitaries, made renewed threats against the Armenian Church.[78] As a consequence, Romanos's downfall is portrayed as a matter of divine retribution, after which the Byzantine emperors largely disappear as active players in the *Chronicle*. However, Andrews has pointed out that Matthew's negative view of Romanos was biased because the author was, in general, hostile towards all the Armenian nobility and Byzantine notables—including emperor Romanos—who were on friendly terms with the Armenian general and *doux* of

75 Andrews, *Matt'eos Urhayec'i*, 75–91.

76 Andrews, *Matt'eos Urhayec'i*, 87–100.

77 Matthew of Edessa, *Chronicle*, 95–96.

78 Robert Bedrosian, ed., *Aristakes Lastivertc'i's History* (New York: Sources of the Armenian Tradition, 1985), 170–72; Matthew of Edessa, *Chronicle*, 133. Repeated in: Robert Bedrosian, ed., *Smbat Sparapet's Chronicle* (Long Branch: Sources of the Armenian Tradition, 2005), 36–37; Bedrosian, ed., *The Chronicle of Michael the Great, Patriarch of the Syrians* (Long Branch: Sources of the Armenian Tradition, 2013), 164.

Antioch Philaretos Brakhamios, himself a political enemy of the Armenian noble family of the Pahlavuni (the *Katholikos* Gregory II in particular) with whom Matthew had very good relations.[79]

There are other primary sources that originate from the historiographical tradition of both Byzantine and non-Chalcedonian Church communities, which can be of great use for the socio-political history of Anatolia and Syria in the eleventh century, but are far less detailed about Romanos's campaigns of 1068–1071. There is Aristakes Lastivertc'i, an eleventh-century Armenian cleric and historian whose *History* (drafted between 1072 and 1079) covers the events of 1000–1071;[80] it includes information on Byzantine-Armenian relations, the Seljuk invasions, and the T'ondrakac'i movement in the Armenian Church. Michael the Syrian, who was a Jacobite patriarch of Antioch (1166–1199) born in Melitene (Malatya) in 1126, who wrote a universal chronicle in Syriac that details events from the Creation until 1195.[81] Michael's *Chronicle* details the succession of empires, with much information concerning the history of Syria, Mesopotamia, and Iran, and the reigns of the Turkmen in Persia, Assyria, Mesopotamia, and Khorasan. Abū al-Faraj (better known as "bar Hebraeus'), another cleric from Melitene (Malatya) born in 1226, was a noted medieval scholar who wrote the *Chronography*, a secular history from the time of creation down to his own day;[82] in large parts of the *Chronography* he summarizes Michael the Syrian's narrative, while he appears to be acquainted with the Persian account of Seljuk origins, the *Maliknāma*.[83] The late thirteenth-century *Chronicle* by Smbat Sparapet is another major source for the history of the Cilician Armenian kingdom, yet roughly three quarters of the work summarizes the *History* of Matthew of Edessa.[84]

79 Andrews, *Matt'eos Urhayec'i*, 94–95.

80 Tara L. Andrews, "Aristakēs Lastivertc'i," in *Encyclopedia of the Medieval Chronicle*, ed. Graeme Dunphy, Cristian Bratu, https://referenceworks.brillonline.com/entries/encyclopedia-of-the-medieval-chronicle/aristakes-lastivertci-SIM_00221?s.num=3, accessed November 3, 2021. See also: Nikolas Matheou, "Situating the History Attributed to Aristakēs Lastiverc'i, 1000–1072: The Empire of New Rome & Caucasia in the Eleventh Century," (PhD diss., University of Oxford: 2018).

81 Amir Harrak, ed., *The Chronicle of Michael the Great* (The Edessa-Aleppo Syriac Codex) (Piscataway: Gorgias, 2019); W. Witakowski, "Syriac Historiographical Sources," in *Byzantines and Crusaders in Non-Greek sources 1025–1204*, ed. M. Whitby (Oxford: Oxford University Press, 2007), 255–58; Florence Jullien, "Michael the Syrian," in *Encyclopaedia Iranica Online*, https://iranicaonline.org/articles/michael-syrian, accessed November 8, 2021.

82 E. A. W. Budge, ed., *The Chronography of Gregory Abū'l-Faraj...Barhebraeus, Being the First Part of His Political History of the World* (London: Oxford University Press, 1932); David Wilmshurst, ed., *Bar Hebraeus, The Ecclesiastical Chronicle: An English Translation* (Piscataway: Gorgias, 2016); Herman G. B. Teule, "Ebn al-ʿEbrī," in *Encyclopaedia Iranica Online*, https://iranicaonline.org/articles/ebn-al-ebri. accessed November 8, 2021.

83 Claude Cahen, "Le Malik-Nameh et l'histoire des origines Seljukides," *Oriens* 2 (1949): 31–65.

84 Tara L. Andrews, "Smbat Sparapet," in *Encyclopedia of the Medieval Chronicle*, ed. Graeme Dunphy, Cristian Bratu (Leiden: Brill, 2010), 1374–75.

The oldest surviving work of eleventh-century Byzantine historiography was probably written by the greatest intellectual of the period, Michael Psellos (b. 1018).[85] Psellos's successful career in the civil service quickly propelled him to the head of the imperial chancery as *protoasecretis* under Constantine IX Monomachos (r. 1042–1055). Shortly thereafter he became not only the trusted advisor of Constantine X Doukas (r. 1059–1067) but also the tutor of his eldest son and heir, the future emperor Michael VII. Yet, Psellos's political star seems to fade away after 1063, following the publication of his *Chronographia*, which saw him "banished" from the court.[86] Nevertheless, Psellos was still regarded as a respected political and military adviser by Romanos IV (r. 1068–71), and he was asked to accompany the emperor in his 1069 campaign against the Turks. Following the publication of a supplement to his *Chronographia* in 1074, his career eventually reached an impasse during the reign of his student, Michael VII (r. 1071–1078), despite the increasingly distorted view of (or, his "subservience to" according to Treadgold) the Doukas family and his student, Michael VII.[87]

Aptly described as a set of imperial biographies, the *Chronographia* deals almost exclusively with the imperial court, yet its treatment of the empire's financial, military, and foreign policies is scanty and impressionistic, hence its limited usefulness as a primary source for the military campaigns of 1068–1071.[88] Psellos looked to paint a portrait of Romanos as a power-hungry rebel in a tyrant's garb, thus prompting Attaleiates's methodical attack on his thesis. Yet, to Psellos's tyrant, Attaleiates counterposed the more subtle image of a man of action,[89] and whilst the *Chronographia* placed the responsibility for the defeat at Manzikert squarely on the shoulders of Romanos, the *History* offers a more analytical description of the battle and a sobre assessment of the reasons (especially the division of the army) that led to the Byzantine defeat.[90] Finally, as Krallis points out, Psellos's encomium of his imperial pupil served as a literary model for Attaleiates's account of the reign of Nikephoros III that may, at first sight, seem to extol the old emperor, yet Krallis emphasized the underlying irony in Attaleiates's description of Nikephoros's policies, an irony that was typical of Psellos's encomia.[91]

85 The starting point for the life and career of Psellos: Stratis Papaioannou, *Michael Psellos: Rhetoric and Authorship in Byzantium* (Cambridge: Cambridge University Press, 2013); Treadgold, *Middle Byzantine Historians*, 272–81; Karpozilos, *Βυζαντινοί Ιστορικοί και Χρονογράφοι*, 3:59–75.

86 The starting point for the *Chronographia*: Efthymia Pietsch, *Die Chronographia des Michael Psellos: Kaisergeschichte, Autobiographie und Apologie* (Wiesbaden: Reichert, 2005); Anthony Kaldellis, *The Argument of Psellos' Chronographia* (Leiden: Brill, 1999); Treadgold, *Middle Byzantine Historians*, 289–308; Karpozilos, *Βυζαντινοί Ιστορικοί και Χρονογράφοι*, 3:75–97; Hunger, *Βυζαντινή λογοτεχνία*, 2:187–201.

87 Treadgold, *Middle Byzantine Historians*, 295.

88 Treadgold, *Middle Byzantine Historians*, 296.

89 Krallis, *Michael Attaleiates*, 84–85.

90 Krallis, *Michael Attaleiates*, 91.

91 Krallis, *Michael Attaleiates*, 113.

The Muslim Sources for the Battle of Manzikert

The primary aim of the following section is to provide an outline of the life and career of the Islamic-educated men who wrote about the Battle of Manzikert. Chase Robinson points out that there is an enormous quantity of "original material" that was available to be "re-produced" by Muslim historians: material that came from a wide-range of government bureaus, both civil and military, in addition to private and semi-private institutions such as mosques, *madrasas*, and libraries.[92] Take, for example, the case of ʿImād al-Dīn al-Iṣfahānī, born into a family of Persian origin that had a tradition of administrative service for the Seljuk dynasty and the Abbasid Caliphate; he became Ṣalāḥ-al-Dīn's secretary, and, except for relatively short periods, was constantly in attendance during his campaigns, dealing with all the manifold chancery tasks, producing letters, appointments, and treaties, and particularly handling the Persian correspondence.[93] Al-Iṣfahānī drew on and quoted a variety of sources for his two-volume "historical prose epic" on Ṣalāḥ-al-Dīn,[94] including diaries, memos, private correspondence, and chancery documents, most often addressed to the caliphal government in Baghdad, drafted by ʿImād al-Dīn on behalf of Ṣalāḥ-al-Dīn or his son and successor al-Afḍal.[95] What is critical, therefore, like the case of their Byzantine counterparts, is the access Muslim historians had to official documents, and this is what I would like to examine here, their sociocultural backgrounds and careers.[96]

Originally from Tortosa in northern Spain, Abū Bakr Muḥammad ibn al-Walīd al-Ṭurṭūshī[97] (ca. 1059–ca. 1126/451–520) emerged as one of the most important Andalusian political philosophers of the Middle Ages. After acquiring an excellent education in law and literature in Zaragoza, al-Ṭurṭūshī eventually settled in Alexandria after 1084, where he taught at a *madrasa* as a leading scholar of jurisprudence and legal theory. Known for his *magnum opus* of Islamic political theory that he addressed to kings and rulers, the *Kitāb Sirāj al-Mulūk* [The Lamp of Kings] was completed in 512/1122. Al-Ṭurṭūshī devoted Chapter 61 of his sixty-four-chapter work to a discussion on different stratagems of war, and it is this chapter that includes not only the earliest extant narrative of the battle but also stories and information from eyewitnesses in circles close to the Seljuk leadership and, possibly, Nizam al-Mulk himself.[98]

---

92 Chase F. Robinson, *Islamic Historiography* (Oxford: Oxford University Press, 2003), 144–45.

93 Lutz Richter-Bernburg, "ʿImād al-Dīn al-Iṣfahānī," in *Medieval Muslim Historians and the Franks in the Levant*, 36.

94 Richter-Bernburg, "ʿImād al-Dīn al-Iṣfahānī," 44.

95 Richter-Bernburg, "ʿImād al-Dīn al-Iṣfahānī," 45; Robinson, *Islamic Historiography*, 145.

96 Robinson, *Islamic Historiography*, chapter 9.

97 Kitāb al-ḥawādiṯ wa-l-bidaʿ, *El libro de las novedades y las innovaciones*, ed. and trans. Ma. Isabel Fierro (Madrid: Consejo Superior de Investigaciones Científicas, 1993); Maximiliano A., *Alarcón y Santón: Lámpara de los príncipes* (Madrid: Instituto de Valencia de Don Juan; Hispanic Society of America, 1930–1931); Hillenbrand, *Turkish Myth*, 26–27; A. Ben Abdesselem, "al-Ṭurṭūs̲h̲ī," in *Encyclopaedia of Islam Online*, https://referenceworks.brillonline.com/entries/encyclopaedia-of-islam-2/al-turtushi-SIM_7650?lang=en, accessed November 8, 2021.

98 Hillenbrand, *Turkish Myth*, 26 (27–29 for the translation of the Manzikert account).

Then, there is the *Dhayl taʾrīkh Dimashq* [Continuation of the History of Damascus], an annalistic chronicle written by Ibn al-Qalānisī[99] (ca. 1073–1160/465–555), a Damascene who was well educated in literature, theology, and law, and who went into public service as head (*ʿamīd*) of the chancery and twice served as mayor (*raʾīs*) of Damascus. Both al-Qalānisī's official positions and his professional connections would have equipped him well for his literary undertaking, since not only he and his father were contemporaries of the events he describes, but he also makes extensive use of eyewitness accounts and documentary evidence.[100] Nevertheless, his Manzikert account has a second-hand feel to it.[101]

A relatively brief but early account of the Battle of Manzikert was inserted into the middle of the *Taʾrīkh Mayyāfāriqīn wa-Āmid* [The History of Mayyafariqin and Amid], a chronicle that focused on the area around Mayyāfāriqīn and Āmid in the Diyār Bakr that was critically close to the battle. It was written by Ibn al-Azraq al-Fāriqī (d. after 1176/572),[102] of Arabic, Kurdish, or Turkish origin born in Mayyāfāriqīn (Gr. *Μαρτυρόπολις*, mod. *Silvan* in Turkey), who made a career in the civil service working for Temürtash, the Artuqid ruler of Mayyāfāriqīn (r. 1124–1154/518–548) and several other Muslim and non-Muslim rulers in eastern Anatolia, Georgia, and Damascus. His career and connections would inevitably have given him access to official records and chancellery documents to use in his work.

Full of interesting details about the Battle of Manzikert is the work of the known historian, qāḍī, and preacher of the Ḥanbalī legal school living in Baghdad, Ibn al-Jawzī (ca. 1116–1201/510–597).[103] For most of his life, Ibn al-Jawzī maintained close relationships with the Caliphs, viziers, and the leading scholars, and his *al-Muntazam fī taʿrīkh al-muluk wa-al-umam* [The Well-Organized (Book) Concerning the History of Kings and Peoples] often is packed with very detailed biographies. Technically, the *al-Muntazam* is a universal chronicle until 1179/574, focusing on the history of Mesopotamia and especially Baghdad. Yet, al-Jawzī's account of Man-

---

99 The following translation does not include the Manzikert account, yet the introduction on the life and work of Ibn al-Qalānisī is good enough. Ibn al-Qalānisī, *The Damascus Chronicle of the Crusades*, trans. by H. A. R. Gibb (London, 1932; repr. 2002); Niall Christie, "Ibn al-Qalānisī," in *Medieval Muslim Historians and the Franks in the Levant*, ed. by Alex Mallett (Leiden: Brill, 2014), 7–28. For the Manzikert account see: Hillenbrand, *Turkish Myth*, 30–31.

100 Gibb, *The Damascus Chronicle*, 9–10.

101 Hillenbrand, *Turkish Myth*, 32.

102 David Thomas and Alexander Mallett, "Ibn Al-Azraq," in *Christian-Muslim Relations: A Bibliographical History, Volume 3 (1050–1200)*, ed. by David Thomas and Alexander Mallett (Leiden: Brill, 2011), 690–94; Chase F. Robinson, "Ibn Al-Azraq, his Ta'rīkh Mayyāfāriqīn, and Early Islam," *Journal of the Royal Asiatic Society* 6 (1996): 7–27; Carole Hillenbrand, "The History of the Jazira 1100–1150: The Contribution of Ibn Al-Azraq al-Fariqi," (PhD diss., University of Edinburgh, 1979). The translation of the Manzikert account: Hillenbrand, *Turkish Myth*, 34–35.

103 Alex Mallett, "Ibn al-Jawzī," in *Christian-Muslim Relations: A Bibliographical History, Volume 3 (1050–1200)*, 731–32; Konrad Hirschler, "Ibn al-Jawzī, ʿAbd al-RaḤmān," in *Encyclopedia of the Medieval Chronicle*, 825. The translation of the Manzikert account: Hillenbrand, *Turkish Myth*, 37–43.

zikert is an unusually long excursus away from his particular area of interest.[104] Drawing heavily from the account of al-Jawzī,[105] although he has been criticized for not citing his sources (except al-Ṭabarī), is one of the most influential historians of the late Middle Ages, Ibn al-Athīr (1160–1233/555–630),[106] whose *al-Kāmil fi'l-ta'rīkh* [The Complete History] aspires to offer an encyclopaedic view in a single text of all the important historical events up to the year 1233/630, with a specific focus on the history of *dār al-Islām*, East and West.

Another interesting account of Manzikert that comes from the second quarter of the thirteenth century contains details not found in earlier accounts (for example, placing Alp Arslan in Khoy, north of Lake Urmia, rather than in Syria, when he heard the news of Romanos's advance to Manzikert) is the *Akhbār al-dawla al-saljūqiyya* [The History of the Seljuq State], penned probably (but not conclusively) by a late twelfth-century Iranian author, al-Ḥusaynī (d. after 1225/622).[107] This is a chronicle of the Great Seljuk Sultanate from its emergence in the Iranian lands in the tenth century until the death of the last Great Seljuk sultan, Ṭog͟hrul III (d. 1194/590). The great value of the *Akhbār* lies in the fact that the narrative focuses on the dynastic, political, and military history of the Seljuks, especially on their efforts to expand the Islamic frontiers in northwestern Iran, Armenia, and the Caucasus, and it also includes detailed sections on the reigns of sultans and on important events, like Alp Arslan's campaigns into Byzantium and the death of Nizam al-Mulk.

A very long narrative that contains rare information about Romanos IV's multi-ethnic army was drafted by al-Bundārī.[108] He was an early thirteenth-century historian who wrote in Arabic and is primarily known for his revision of the *History of the Seljukids* by ʿImād al-Dīn al-Iṣfahānī, a work al-Bundārī dedicated to al-Muʿaẓẓam Isa, the Ayyubid emir of Damascus, in 1226/623. However, as useful as this source may be, al-Bundārī muffles the defection of Turkish contingents in Romanos's army to Alp Arslan's side during the battle.

Another full account of Manzikert that offers significant details about Alp Arslan's activities leading up to the battle itself, including a rare and thorough description of

104 Hillenbrand, *Turkish Myth*, 37.

105 Hillenbrand, *Turkish Myth*, 67.

106 Françoise Micheau, "Ibn al-Athīr," in *Medieval Muslim Historians and the Franks in the Levant*, ed. Alex Mallett (Leiden: Brill, 2014), 52–83; David Waines and Heidi R. Krauss-Sánchez, "Ibn al-'Athīr," in *Encyclopedia of the Medieval Chronicle*, 823; David S. Richards, *The Chronicle of Ibn al-Athir for the Crusading Period from al-Kamil fi'l-Ta"rikh, Part 1: The Years 491–541/1097–1146: The Coming of the Franks and the Muslim Response* (London: Routledge, 2006), 1–10. The translation of the Manzikert account: Hillenbrand, *Turkish Myth*, 64–66.

107 Clifford Edmund Bosworth, *The History of the Seljuq State: A Translation with Commentary of the Akhbār aldawla al-saljūqiyya* (London: Routledge, 2011); David Durand-Guédy, "al-Ḥusaynī, Ṣadr al-Dīn," in *Encyclopaedia of Islam Online*, http://dx.doi.org/10.1163/1573-3912_ei3_COM_30591, accessed November 9, 2021

108 M. Th. Houtsma, and Cl. Cahen, "al-Bundārī," in *Encyclopaedia of Islam Online*, http://dx.doi.org/10.1163/1573-3912_islam_SIM_1525, accessed November 9, 2021. The translation of the Manzikert account: Hillenbrand, *Turkish Myth*, 58–63, accessed November 9, 2021.

the Byzantine army's impedimenta for the 1071 campaign, was penned by a renowned preacher and historian, Sibṭ Ibn al-Jawzī (1185–1256/581–654).[109] His *Mirʾāt al-zamān fī ta ʾrīkh al-aʿyān* is a universal history up to the year 1256/654 and a biographical dictionary that focuses on the most important members of Sibṭ Ibn al-Jawzī's own professional circle of the *ulamā* and the Ayyūbid leaders of Damascus. Although his original material for the period before his own lifetime is relatively limited, he relies on a wider range of sources for the eleventh and twelfth centuries and, particularly for Syria and Egypt, he worked with the chronicles of Ibn al-Qalānisī, Ibn al-Athīr, and Ibn al-Azraq al-Fāriqī.

One of the most useful primary sources from the thirteenth century was penned by Ibn al-ʿAdīm, a jurist and historian born in Aleppo (1192–1262/588–660),[110] who rose to prominence as a professor of *fiqh* (Arabic: "understanding"; Muslim jurisprudence) in two Ḥanafī madrasas in Aleppo between 1220/616 and 1236/633. As a diplomat, al-ʿAdīm played a leading role in the service of the Ayyūbid rulers of Aleppo and Damascus, and was charged with many missions and ambassadorial roles that gave him ample opportunity to travel; it was during these travels he visited many scholars where he gained much of the information he would later use in his writings. He wrote the *Bughyat al-ṭalab fī taʾrīkh Ḥalab* [short: *Taʾrīkh Ḥalab*, The History of Aleppo], a multi-volume collection of biographies of famous men from Aleppo, and the *Zubdat al-ḥalab min taʾrīkh Ḥalab* [The Cream of the Milk from the History of Aleppo], one of the richest histories of Syria during the period of the crusades from early times until 1243/641. The in particular *Zubdat* contains a rare attempt to make sense of the military tactics at Manzikert, with special attention to the nomadic tactics of the Seljuks, while also stressing the multi-ethnic nature of Romanos's army, although the author avoids any mention of the defection of some of these troops to Alp Arslan's army on the eve of the battle.

Finally, Muḥammad Ibn Khāvandshāh ibn Maḥmūd, better known as Mīrkhwānd (b. 1433/836, Balkh–d. 1498/903, Herāt),[111] was one of the most important chroniclers of Persia who spent most of his life in Herāt, in the court of the last Timurid Sultan, Ḥusayn

---

109 Alex Mallett, "Sibṭ Ibn al-Jawzī," in *Medieval Muslim Historians and the Franks in the Levant*, 84–108; Heidi R. Krauss-Sánchez, "Sibṭ Ibn al-Jawzī," in *Encyclopedia of the Medieval Chronicle*, 1356–57. The translation of the Manzikert account: Hillenbrand, *Turkish Myth*, 67–73.

110 Anne-Marie Eddé, "Kamāl al-Dīn ʿUmar Ibn al-ʿAdīm," in *Medieval Muslim Historians and the Franks in the Levant*, 109–34; Konrad Hirschler, "Ibn al-Adim, Kamal al-Din ʿUmar," in *Encyclopedia of the Medieval Chronicle*, 822; B. Lewis, "Ibn al-ʿAdīm," in *Encyclopaedia of Islam Online*, http://dx.doi.org/10.1163/1573-3912_islam_SIM_3063, accessed November 15, 2021. For the life and milieu of al-ʿAdīm, as reflected in his Bughyat: David W. Morray, *An Ayyubid Notable and His World: Ibn Al-ʿAdīm and Aleppo as Portrayed in His Biographical Dictionary of People Associated with the City* (Leiden: Brill, 1994). The translation of the Manzikert account: Hillenbrand, *Turkish Myth*, 74–77.

111 Jan Schmidt, "Mirkhwand," in *Encyclopedia of the Medieval Chronicle*, 1113; A. Beveridge and Beatrice Forbes Manz, "Mīrk͟hwānd," in *Encyclopaedia of Islam Online*, (http://dx.doi.org/10.1163/1573-3912_islam_SIM_5224, accessed November 15, 2021. The translation of the Manzikert account: Hillenbrand, *Turkish Myth*, 98–104.

Bayqarah (1469–1506). At the request of his patron, he compiled his universal history in Persian, the *Rawḍat al-Ṣafā' fī sīrat al-anbiyā' wa 'l-mulūk wa 'l-k͟hulafā'* [The Garden of Purity: Respecting the Lives of the Prophets, the Kings and the Caliphs]. The work is composed of seven large volumes and a geographic appendix, and it begins with the age of the pre-Islāmic Persian kings and surveys the major Muslim rulers of Iran up to the events of 1523 (completed by the historian's son-in-law, Khwandamir). Although Mīrkhwānd's account provides little new details on the Battle of Manzikert, it offers a rare glimpse into Romanos's alleged strategic ambitions to take Baghdad and march on to Samarkand.

Chapter 2

CHRISTIAN AND MUSLIM SOURCES FOR THE BATTLE OF MANZIKERT

MAKING SENSE OF THE "BATTLE-PIECE"

THE MAIN AIM of this chapter is to examine the primary sources for the Battle of Manzikert strictly from a military perspective and attempt to reach some conclusions regarding their value for the history of eleventh-century warfare in the region of Asia Minor. The major questions that will be raised in this chapter are: how accurate and detailed are their descriptions of pitched battles and clashes between enemy units during the Byzantine campaigns of 1068 to 1071 in the East? How far do their narratives permit an accurate reconstruction of chains of events, especially regarding the battlefield manoeuvres of armies in action?

The "Battle-Piece"

Why do historians write "battle-pieces"? And how should a modern reader attempt to understand the narrative of such a dramatic and complex event? Examining what a battle narrative can tell us and how to read it should be a central part of our approach to military history. But first, what is a "battle-piece"? We could say that it is the battle account that forms the essential element in the warfare narrative; a literary piece that was incorporated into hagiographies, letters, rhetorical works, epic poetry, romances, and, of course, historical texts.[1] Yet, what remains unclear in the aforementioned definition is the perspective from which the battle narrative is written. Because the vast majority of the battle narratives from ancient to early modern times focus on events from the perspective of the commander(s) who inevitably had a disproportionate influence on the course of events, modern scholars have followed the same well-trodden course of writing about battles, adopting a traditional "command-centred" approach to the strategic, operational, and tactical movements of armies, where the thoughts, decisions, and actions of the commander(s) are far more visible than those of the common soldier(s).

Accordingly, one of the toughest challenges involved in writing about medieval warfare is describing what battle in that period "was really like": how it might have felt to be there, including not just the physical but also the emotional sensations. What would have been running through the combatants' minds? What were the motivations used to suppress their primal instinct for running away from danger? The issue of battlefield emotions has been hugely neglected to the point of being considered a "taboo" until the very end of the twentieth century. And it was in breaking this "taboo" that military histo-

1 McGrath, "Warfare as Literary Narrative," 160–61 (especially nos. 1 and 2).

rians owe immense gratitude to John Keegan who, in his 1976 classic *The Face of Battle*, takes a look at the direct experience of individuals in three major encounters of the last six hundred years: Agincourt (1415), Waterloo (1815), and the Somme (1916) at "the point of maximum danger." Keegan's legacy continues to inspire young scholars in the field several decades later, with Łukasz Różycki tackling the issue of battlefield emotions, including fear and post-traumatic stress, of Roman soldiers in Late Antiquity, and the methods employed to harness them, while Conor Whately reconstructs an image of traditional aspects of the Byzantine army, such as unit types and recruitment, though the study of the ordinary soldier of the sixth century.[2]

Keegan's approach to battle narrative has been dubbed the "face of battle" narrative for reasons that merit no explanation. This approach aims to understand the battle "as it really was"; the physical environment, human psyche, blood and guts, and all the "small group dynamics" in one historical narrative that describes this complex event from the perspective of the ordinary soldier, while trying to provide answers to questions about leadership, motivation, discipline, courage, fear, and other emotions. The "face of battle" approach, therefore, may come some way in addressing the aforementioned imbalance in the focus of the military historiography because, in Keegan's words, the "exploration of the combatants' emotions, if not the indulgence of our own, is essential to the truthful writing of military history."[3]

In her masterful analysis of the "battle-pieces" of Ammianus Marcellinus and Julius Caesar, Kimberly Kagan criticizes Keegan's historical philosophy and argues that the "face of battle" narrative alone actually obscures the critical events of the battle, thus hampering the historian's effort to look for causal events to explain their outcomes. One can see this major drawback in Ammianus's narratives for the Sassanid siege of Amida in 359, a military event that had the most important formative effect on his understanding of war and the nature of combat after he became personally involved. Yet, as a middle-ranking participant, Ammianus found it difficult to identify the causal chain of different battle episodes, offering instead a version of events that made an impression on him–the "Ammianus version of events."[4]

Worst still, as Kagan explains, Ammianus faced his biggest narrative challenge when he wished to describe a battle he did not experience—the Battle of Strasburg in 357; Ammianus made use of his experience in the static siege operation of Amida to try to describe how he imagined the nature of combat at Strasburg would have unfolded, an experience (or lack of) that would also have reinforced Ammianus's distorted impression of Emperor Julian as a static commander.[5] The "face of battle" approach, therefore, vividly conveys how the battle felt, sounded and looked, but, compared to the traditional

---

2 Both works came out in the first half of 2021: Łukasz Różycki, *Battlefield Emotions in Late Antiquity: A Study of Fear and Motivation in Roman Military Treatises* (Leiden: Brill, 2021); Conor Whately, *Procopius on Soldiers and Military Institutions in the Sixth-Century Roman Empire* (Leiden: Brill, 2021).

3 Keegan, *The Face of Battle*, 16.

4 Kimberly Kagan, *The Eye of Command* (Ann Arbor: University of Michigan Press, 2006), 23–51.

5 Kimberly Kagan, *The Eye of Command*, 52–95, 192.

"command-centred" approach, it fails to explore the decision-making process and to provide insights that could explain the outcome of such a notoriously complex and non-linear event.

The inevitable comparison here is between the "Ammianus version of events" for Strasburg and the "Attaleiates version of events" for Manzikert. In the tradition of Thucydides and Polybius, Attaleiates underlines several times that he relied on what "I witnessed myself first-hand, not what I learned by hearsay and the unreliable reports of others."[6] In fact, we saw that what distinguishes Attaleiates from all the other sources for the Battle of Manzikert is his participation in a number of the campaigns he describes. Yet, we need to be mindful of the fact that Attaleiates was a participant but not a combatant in the campaigns led by Romanos between 1068 and 1071.[7] We see this in his description of the clash outside Hierapolis in northern Syria in mid-November 1068, where he notes: "It was then that I not so much despaired of my own safety as I came to despise the cowardice, ineptitude, or wretchedness of the Romans. For although the Romans had suffered an overwhelming defeat before their encampment, none of the remaining companies or officers was moved to action."[8] Apparently, at Hierapolis, Attaleiates would have been a mere observer and his field of vision could not have been much clearer than that of a mere soldier, which takes us back to the "face of battle" approach we saw earlier.[9] Indeed, how much can we rely on Attaleiates's scorn of the tactical reaction of the cavalry soldiers of the tagma of the *Scholai* to the attacks of Turkmen horse-archers, when: "even while witnessing the defeat of the others [*Stratelatai*], gave no thought to coming to their assistance or doing anything else expected of soldiers, but rather they stayed fixed there, apparently with the idea that they would not be noticed by the enemy if they stood calmly at ease."[10] I will show further down in this chapter that the Byzantines were highly experienced in their dealings with lightly armed and mobile cavalry units firing arrows or javelins at them from a distance, and the typical tactic of a much more heavily armed cavalry unit was to seek shelter inside a hollow infantry square or to form a compact unit, and certainly not to break formation in pursuit of a more mobile enemy or to help their comrades; Attaleiates apparently did not fully grasp the tactical situation!

Accordingly, one can discern similar restrictions in Attaleiates's perspective of the fateful campaign at Manzikert in 1071. He vividly narrates the apparent confusion in the imperial camp caused by the concerted attacks of bands of Turkish light cavalry on those Byzantines gathering supplies. Following the failed attempts by Bryennius and Basilakes to collect intelligence about the tactical situation, Attaleiates puts the emperor leaving the camp to investigate further. But since he merely reports that the army returned at

6 Attaleiates, *History*, 11. See also: *History*, 7, 189, 285, 359, 467, 587. Attaleiates's historical method with its emphasis on the study of causation may have its roots in Psellos's lectures on science and nature: Krallis, *Michael Attaleiates*, 64–65, 74.

7 Attaleiates's job as a judge and a "legal expert/adviser" is clear. Attaleiates, *History*, 11.

8 Attaleiates, *History*, 207.

9 Sinclair, "War Writing," 82–84.

10 Attaleiates, *History*, 205.

sunset, he probably did not accompany the party. Instead, he refers to the Turks falling upon the unsuspected Patzinaks roaming outside the camp "as if *ex machina*," again being unable to grasp the Turkish tactic of attrition and surprise attacks.[11] Attaleiates describes the flow of wounded men in the camp and the psychological impact ("they were groaning from the pain of their wounds") on the men who stayed behind, while he vividly describes the sleepless night the camp endured as a result of the "inarticulate howling and shooting of arrows they [Turks] rode around terrifying them and placing them in a tight spot."[12]

Attaleiates's account perfectly conveys the sense of panic and confusion among the imperial army as the battle unfolded, while it is clear that he relied on the testimony of others for the events beyond the camp.[13] For Andronikos Doukas's retreat from the battlefield, Attaleiates relies on "ὡς δ' οἱ πολλοί πληροφοροῦσιν" (on what many have related).[14] As the army poured back into the camp, he deplores the lack of an "informed statement," writing that: "when we asked them what happened, they replied that they had not seen the emperor. It was like an earthquake with howling, grief, sudden fear, clouds of dust, and, finally, hordes of Turks riding all around."[15] Finally, when it came to the fate of the emperor, he mentions that 'some claimed that the emperor was firmly resisting with what was left of his army and that he had routed the barbarians. Others said that he had been killed or captured," suggesting further down that victory changed hands until many of the Kappadokians standing with the emperor began to flee and whom he, personally, tried to send back to their posts.[16] To reiterate, therefore, that Attaleiates's view at Manzikert was that of a camp-follower, able to notice events in and around the camp, but only able to rely on hearsay for what was happening further afield and, most likely, he tried to make sense of the course of the battle when he was back in the safety of Constantinople; hence, what we should call the "Attaleiates version of events" for Manzikert.[17]

This leads us to the next two questions: why did medieval historians write battle narratives, and what purpose did these serve? The short answer for both is: for reasons that differ fundamentally from what we (modern historians) consider to be good, detailed, and objective history. Yet, it would certainly be incorrect and unproductive to assess

---

11 Attaleiates, *History*, 279–85.

12 Attaleiates, *History*, 285.

13 Sinclair, "War Writing," 84–86. An excellent re-evaluation of Attaleiates's role in the battle: Antonios Vratimos, "Was *Michael Attaleiates* present at the battle of Mantzikert?," *Byzantinische Zeitschrift* 105 (2012): 829–40.

14 Attaleiates, *History*, 292–93.

15 Attaleiates, *History*, 295.

16 Attaleiates, *History*, 295.

17 About "Procopius's version of events," and how his view of the episodes during the Gothic siege of Rome between 537 and 538 was muddled by the landscape. Christopher Lillington-Martin, "Procopius on the Struggle for Dara in 530 and Rome in 537–538: Reconciling Texts and Landscapes," in *War and Warfare in Late Antiquity*, ed. Alexander Sarantis and Neil Christie (Leiden: Brill, 2013), 599–630.

and explain the descriptions of battle in the narratives of Byzantine historians with a view towards whether they heed what we think a (military) historian should examine; rather, we must be mindful of their cultural milieu, and what factors seemed decisive, and what events owed to be saved for posterity in the eyes of the historians.[18] As Byzantine classicizing historians produced their histories in the grand manner of Herodotus, Thucydides, and Polybius, this meant they had to meet several for their works to be regarded as authoritative and, thus, survive: the account would be truthful and generally devoid of bias, and his/her work would have some moral and/or didactic value; the narrator would be omniscient; the narrative would be full of ethnographic digressions; and the narrative would be full of battles and sieges, often preceded by speeches and even letters.[19]

A Truthful Account of Moral and/or Didactic Value

First, the historian would, as he would proclaim in his/her *prooimion*,[20] be writing an account which is truthful and generally devoid of bias, while his/her work would have some moral and/or didactic value for future readers. Of all the claims made by ancient and medieval historians, the assurance that it was uppermost in their mind to be impartial is by far the most common, topped up with another commonplace *topos* of criticizing their predecessors for flattery and bias because they counted on some sort of material reward or because of personal grudges.[21] That was what the ancient historians disapproved of the most—the universal tendency to take sides, and its rejection, therefore, became a fundamental element in determining the historian's authority. Yet, because the historians also wanted to be accurate in the description and evaluation of people's deeds so that audiences could draw the proper moral lessons from history, they inevitably had to offer words of praise or blame. But this is not the polar opposite of being truthful; imaginative reconstruction (*εὕρεσις-εὑρίσκω*; *inventio*) of deeds was common, but making up (*ἐπιψεύδομαι*; *ἀναπλάσσω*) events in order to fit the historian's "agenda" was a cardinal sin.[22]

---

18 Conor Whately, *Battles and Generals: Combat, Culture, and Didacticism in Procopius' Wars* (Leiden: Brill, 2016), 3–37.

19 Whately, *Combat, Culture, and Didacticism*, 1 and 223; Anthony Kaldellis, *Ethnography After Antiquity: Foreign Lands and Peoples in Byzantine Literature* (Philadelphia: University of Pennsylvania Press, 2013), 1–2. An authoritative work for ancient historiography is by John Marincola, *Authority and Tradition in Ancient Historiography* (Cambridge: Cambridge University Press, 1997).

20 Iordanis Grigoriadis, "A Study of the Prooimion of Zonaras' Chronicle in Relation to Other 12th-century Prooimia," *Byzantinische Zeitschrift* 91 (1998): 327–44.

21 Marincola, *Authority and Tradition*, 158–59 (claims of impartiality), 218–36 (polemic against predecessors); Charles W. Fornara, *The Nature of History in Ancient Greece and Rome* (Berkeley: University of California Press, 1983), 91–141; Bernard Lewis, *History: Remembered, Recovered, Invented* (Princeton: Princeton University Press, 1975).

22 Marincola, *Authority and Tradition*, 160–62.

As with ancient historiography, the Byzantine concern for truth was often about avoiding flattery or false condemnation, very often coupled with blaming (or complaining about) one's predecessors for similar lapses in their historical methodology. Polybius perceived his role as an impartial historian as being able to praise his enemies and censure his kinsmen where necessary, a view repeated almost verbatim by Anna Komnene many centuries later.[23] At the beginning of *The Wars of the Jews*, Flavius Josephus also berates those who have repeated fabricated histories "either out of a humour of flattery to the Romans, or of hatred towards the Jews."[24] Again in the same fashion, Anna underscores her historical veracity by repeating that "I am by nature averse to fabricating tales and inventing slanders, though I know this is a common practice, especially if people are bitten by envy or malice, nor do I lend a ready ear to popular calumnies."[25] Despite Attaleiates's "agenda" that we saw in the previous chapter, he feels compelled to remind his readers that he also tried to be balanced in "setting out the virtues and vices of the rulers and the other men in power."[26] On top of this, when he feels that his story about the rise of his "hero" Botaneiates to the throne may seem implausible to his readers, he adds: "I think, however, that some in our audience may argue that all this has been written by me to praise and not only to narrate, that I am allegedly currying favor with the ruler, given that past emperors also honoured and rewarded with gifts the loyalty of their subjects. But if they think this, they have not grasped the matter well."[27] Finally, Skylitzes lays out his historiographical method in plain terms in the foreword to his Synopsis: "I read the histories of the abovementioned writers with great care. I conjured away from them all comments of a subjective or fanciful nature. I left aside the writers' differences and contradictions. I excised whatever I found there which tended toward fantasy; but I garnered whatever seemed likely and not beyond the bounds of credibility."[28]

Similar statements of seeking the truth and applying critical thinking in their writings were common amongst Muslim historiographers as well.[29] Yet, these statements were, mostly, stereotypical, or, just a "statement of will," as the majority were well aware of the contradictory evidence, fabrications/propaganda, exaggerations and bias that rendered "objective history" almost an impossibility. We read about Ibn al-Qalānisī's historiographical methods in an excursus under date A.H. 540:

> I have completed the narrative of events set forth in this chronicle, and I have arranged them in order and taken precautions against error and rashness of judgment and careless slips in the materials which I have transcribed from the

23 Polybius, *Histories*, 1;14; *Alexiad*, preface:2.

24 Flavius Josephus, *The Wars of the Jews*, 1:prologue.

25 Komnena, *Alexiad*, bk. 3, 1.

26 Attaleiates, *History*, 7.

27 Attaleiates, *History*, 513. See also Attaleiates, *History*, 587.

28 Skylitzes, *Synopsis*, 3.

29 Robinson, *Islamic Historiography*, 143–55.

> mouths of trustworthy persons and have transmitted after exerting myself to make the fullest investigations so as to verify them, down to this blessed year 540. Since the year 535 and down to this point, I had been engaged with matters which distracted my mind from making the fullest enquiries into those current events which required to be set down in this book, and from seeking out the truth concerning them and all the attendant circumstances. Consequently, I left a blank space after the events of each year, in order to insert therein those narratives and events the truth of which was ascertained.[30]

Nonetheless, if we look at the "historical record" of ʿImād al-Dīn al-Iṣfahānī's portrayal of Ṣalāḥ-al-Dīn, he is clearly "economical" with the truth, as he only allows himself to criticize the sultan's entourage but never the sultan himself, whose unwavering resolve he continually emphasizes.[31]

The enduring belief amongst ancient and medieval historians that their works would have real practical value for posterity can explain one of the most familiar *topoi* that can be found in the prefaces to these works—the emphasis on the moral/didactic value of their text.[32] The model for all future historians, Thucydides, for instance, aspired for his *History* to "be judged useful by those inquirers who desire an exact knowledge of the past as an aid to the interpretation of the future."[33] In fact, de Romilly went as far as to argue that Thucydides shaped his battle descriptions as lessons for future generals.[34] Polybius also adopted this practice and used the wars against the Gauls or Celts as a valuable case in point to showcase the Roman military progress and adaptability in the Gallic operational theatre, for he too believed that the purpose of his *History* was, among other things, to be a teacher of character and to provide lessons for future political and military leaders.[35] Following in the footsteps of these "giants," Procopius also shares the view of the didactic quality of histories, something which he makes clear just a few lines into his *Wars* when he notes down his wish for his work to serve as a repository of knowledge "for men who purpose to enter upon

30 Ibn al-Qalānisī, *The Damascus Chronicle of the Crusades*, 10.

31 Richter-Bernburg, "ʿImād al-Dīn al-Iṣfahānī," 46.

32 Whately, *Combat, Culture, and Didacticism*, 2–6, 224–31; Neville, *Guide to Byzantine Historical Writing*, 8–25; Neville, *Heroes and Romans*, 29–38, 194–203; Neville, "Why Did the Byzantines Write History?," in *Proceedings of the 23rd International Congress of Byzantine Studies Belgrade, 22–27 August 2016*, ed. Smilja Marjanović-Dušanić (Belgrade: The Serbian National Committee of AIEB, 2016), 265–76; Warren Treadgold, "The Unwritten Rules for Writing Byzantine History," in *Proceedings of the 23rd International Congress of Byzantine Studies Belgrade*, 277–92; Athanasios Markopoulos, "Byzantine History Writing at the End of the First Millennium," in *Byzantium in the Year 1000*, ed. Paul Magdalino (Leiden: Brill, 2003), 183–97; Robinson, *Islamic Historiography*, particularly chapters 4 and 5, and the conclusions. For the didactic significance of various landmarks in the career of Nur al-Din in the Muslim sources: Hillenbrand, *The Crusades*, 132–41.

33 Thucydides, *The Peloponnesian War*, bk. 1, chap. 22, 4.

34 Jacqueline de Romilly, *Histoire et raison chez Thucydide* (Paris: Belles-Lettres, 1956), 153.

35 Polybius, *Histories*, bk. 2, 35; bk. 12, 25. See also: Daniel Walker Moore, "Learning from Experience: Polybius and the Progress of Rome," *The Classical Quarterly* 67 (2017): 132–48.

a war or are preparing themselves for any kind of struggle."[36] I will show further how the histories and the military manuals of Late Antique Rome and Byzantium, which constitute a distinct literary genre, more or less complemented each other in the eyes of their audience and both should be studied in parallel for their advisory nature.

In his opening chapter, Leo the Deacon asserts the "useful and profitable" aspects of history: it recounts deeds, provides examples "to emulate...and avoid," and "is agreed to be as profitable as the other useful things in life, inasmuch as it brings mortal affairs back to life or gives them youthful vigour."[37] Skylitzes also declares from the outset in his *Synopsis* that he intends to provide a concise history that would be useful *τοῖς φιλιστοροῦσιν* ("to those who love history") and would serve as a manual for his audience "to take along with them and consult as a handbook."[38] Likewise, Attaleiates highlights that history "has proven to be exceedingly useful for life, as it reveals the lives of those who were virtuous and those who were not, describes illustrious deeds born of flawless planning and effort as well as inglorious actions caused by the faulty planning or negligence of those governing public affairs."[39] For Attaleiates, the whole point of writing history is to "convey clear instruction and set patterns for the future." Therefore, historians become arbiters of morality and character, thus having a similar function to hagiographies and homilies, which also commemorate great deeds of the past and call on their audience to draw moral lessons from the theme being treated.[40]

A good example of Byzantine historians offering their audience divergent moral lessons about the same people can be found in the works of Anna Komnene and Nikephoros Bryennios. They essentially write the same stories about Alexios Komnenos, with Bryennios portraying the future emperor as a great strategist, but one who is also cunning and "un-Roman," while Anna tells essentially the same stories in such a way that her father emerges as the protagonist of the events and a brilliant strategist in the ancient Greek style.[41] A similar "Achilles ethos" versus "Odysseus ethos" dichotomy can be discerned in Procopius's *Vandal Wars*, in which Totila embodies the traditional warrior ideal (i.e., Achilles) whilst Belisarius emerges as another

36 Procopius, *History of the Wars*, bk. 1, chap. 1, 1; bk. 1, chap. 1, 2.

37 *The History of Leo the Deacon: Byzantine Military Expansion in the Tenth Century*, trans. D. Sullivan and A-M. Talbot (Washington DC: Dumbarton Oaks, 2005), 55.

38 Skylitzes, *Synopsis*, 3.

39 Attaleiates, *History*, 9.

40 Neville, "Why Did the Byzantines Write History?," 269; Mary Cunningham, "Homilies," in *The Oxford Handbook of Byzantine Studies*, ed. Elizabeth Jeffreys, with John Haldon and Robin Cormack (Oxford: Oxford University Press, 2008), 862–71; Alice-Mary Talbot, "Hagiography," in *The Oxford Handbook of Byzantine Studies*, 862–71. See also: Anna Lampadaridi, Vincent Déroche, and Christian Høgel (eds.), *L'histoire comme elle se présentait dans l'hagiographie byzantine et médiévale* (Uppsala: Acta Universitatis Upsaliensis, 2022).

41 Military engagements in the Hyle become a form of a "moral theatre" where Bryennios depicts a hero's tactic as either "straightforward" or "devious" (i.e., "honourable" or "cunning"), Neville, *Heroes and Romans*, 89–103, 182–93.

"Odysseus" known for his "intelligence, cleverness, and trickery instead of brute force and open confrontation."[42]

We saw in the previous chapter that Krallis calls Attaleiates's *History* a "point-by-point refutation" of Psellos's (pro-Doukas) account of the military career and blinding of Diogenes (his "second hero").[43] Michael aims to explain and justify Botaneiates's ascension to the throne, without seeking any moral or material reward, and, therefore, "I diligently seek to represent it [Botaneiates's career] clearly for everyone now and to immortalise him as the most beautiful model for future generations who may read my work and remember him."[44] Skylitzes also criticizes Psellos in much harsher terms than Attaleiates for his scheming against poor Diogenes, his determination to diminish the latter's reputation, and his influence on the soft and incompetent Michael VII.[45] Likewise, the *Continuation* includes unflattering details Attaleiates chose to omit about the personal life of Botaneiates after he became emperor, whilst Skylitzes also attempts to "re-fashion" the material from the *History* to rehabilitate the precursor of the Komnenian dynasty, Isaac Komnenos, because the latter's bloody coup *d'état*, his confiscation of ecclesiastical vessels, and his early military disasters made for "uncomfortable parallels" to Alexios's rise to the throne twenty-five years later.[46] On the other hand, Bryennios paints a much less complimentary picture of Diogenes as an emperor and leader, accusing him of being prone to flattery, a victim of circumstances and his poor strategic judgement, whilst he meticulously muffles the critical role of Andronikos Doukas and Joseph Trachaneiotes at Manzikert.[47] In fact, he claims that although Andronikos was indeed hostile to the emperor,[48] he had no involvement in the conviction and blinding of Diogenes the following year (1072), which, on the contrary, he tried to prevent.[49] It is obvious that Bryennios could not criticize or write anything against the family and predecessors (both the Doukai and the Komnenoi) of his patron, the empress Eirene Doukaina, or his ancestor Nikephoros Bryennios ("the Elder").[50]

---

42 Whately, *Combat, Culture, and Didacticism*, 190–96.

43 Krallis, *Michael Attaleiates*, 80–100. See also: Dimitri Krallis, "Attaleiates as a Reader of Psellos," in *Reading Michael Psellos*, ed. Charles E. Barber and David Jenkins (Leiden: Brill, 2016), 167–91.

44 Attaleiates, *History*, 513; Karpozilos, *Βυζαατινοί Ιστορικοί και Χρονογράφοι*, 3:206–8.

45 Skylitzes, *Continuation*, 137.

46 McGeer, *Continuation*, 14–26; Karpozilos, *Βυζαντινοί Ιστορικοί και Χρονογράφοι*, 3:313–17.

47 Bryennios, *Ύλη Ιστορίας*, 78–93; Karpozilos, *Βυζαντινοί Ιστορικοί και Χρονογράφοι*, 3:360–65.

48 "He [Andronikos] was not very friendly to the emperor" is, surely, an understatement: Bryennios, *Ύλη Ιστορίας*, 91.

49 Bryennios, *Ύλη Ιστορίας*, 113–15.

50 Karpozilos, *Βυζαντινοί Ιστορικοί και Χρονογράφοι*, 3:383–88.

An Omniscient Historian: A Guide and a Narrator

Second, as the interaction between the narrator and the audience is essential in a didactic work like history, the historian is bound to be omniscient and to intervene in his/her narrative not only to state his qualifications but to act as a guide and narrator to events. This type of authorial interjection proliferated in the works of classical historians like Herodotus, Thucydides, and Xenophon, with Gribble explaining that "the emergence of the voice of the narrator marks climaxes of the emotional and analytical superstructure of the narrative, and from this arises their structural function, not only to signal culminations, but also to suggest relationships with other similar or related moments in the narrative."[51] Whately offers a rare analysis of Procopius's use of a great number of both explicit and implicit interventions in his "battle-pieces" in the *Vandal Wars*, more so than in the rest of the work, pointing out their fundamental role as a literary tool "in highlighting the importance of the events, and in assisting the readers' interpretation of them."[52]

Roger Scott has argued that the primary difference between classical authors and their Byzantine counterparts is that the former "does not have a personal stake in the subject matter, or, if he does, it only comes in on the fringes of his work,"[53] thus underlining the responsibility (or burden?) of later Byzantine authors to defend narrative choices because of their subject matter that was directly connected to themselves. However, he dates the appearance of this fundamentally different approach sometime after the seventh century, using the *Alexiad* as his foremost example. On this matter, Alexander Kazhdan and Ann Wharton Epstein maintain that until the ninth and tenth centuries, historians were eager to appear objective by minimizing their direct involvement in their narratives and personal connections to their subjects—something akin to a deliberate rejection of their "authorial intrusion"; this will change at the beginning of the eleventh century, with Psellos representing a transition in this technique by incorporating a personal element into his histories, which will culminate in the histories of Gregoras and Kantakouzenus in the fourteenth century.[54] Finally, in an article titled

---

51 David Gribble, "Narrator Interventions in Thucydides," *Journal of Hellenic Studies* 118 (1998): 41–67 (quotation from 62). See also: Matthew Fox and Niall Livingstone, "Rhetoric and Historiography," in *A Companion to Greek Rhetoric*, ed. Ian Worthington (Oxford: Blackwell, 2007), 542–61; Lisa Kallet, "Thucydides' Workshop of History and Utility Outside the Text," in *Brill's Companion to Thucydides*, ed. Antonios Rengakos and Antonios Tsakmakis (Leiden: Brill, 2006), 360–61; Tim Rood, "Objectivity and Authority: Thucydides' Historical Method," in *Brill's Companion to Thucydides*, 225–50; Marincola, *Authority and Tradition*, 182–206; Marincola, *Greek Historians* (Cambridge: Cambridge University Press, 2001), particularly chapters 2, 3, and 5.

52 Whately, *Combat, Culture, and Didacticism*, 122–25.

53 Roger Scott, "The Classical Tradition in Byzantine Historiography," in *Byzantium and the Classical Tradition: University of Birmingham, Thirteenth Spring Symposium of Byzantine Studies, 1979*, ed. Margaret Mullet and Roger Scott (Birmingham: University of Birmingham, 1981), 63.

54 Alexander Petrovich Kazhdan and Ann Wharton Epstein, *Change in Byzantine Culture in the Eleventh and Twelfth Centuries* (Berkeley: University of California Press, 1990), 204–6, 220–23; Alexander Kazhdan, "L'Histoire de Cantacuzène en tant qu'oeuvre littéraire," *Byzantion* 50 (1980): 279–335 (especially 284–85).

"'Writers' Intrusion' in Early Byzantine Literature," Jakov Ljubarskij produces four sets of criteria which can be used for recognizing and categorizing an author's intrusion in historical writing: "immediate self-presentation of the author, direct comment on events and persons (deviating from the narrative), emotional evaluation of events and persons (epithets and adverbs), and references to sources (author's interest in trustfulness)."[55]

In what is probably the most influential contribution to the study of "authorial intrusion" in Byzantine historiography, Ruth Macrides notes that "if students of Thucydides are now looking for the real Thucydides in his silences, Byzantinists could start with the opposite task—to study their authors' explicit intrusions."[56] She argues that the Byzantine historians' "authorial interjections" rarely served the purpose of presenting the audience with his/her credentials for writing a history. For example, Attaleiates and Bryennios both wished to emphasize their "reading of the signs" of the impending defeat and criticize Romanos's rejection of "sound" strategic advice, which reads like the proverbial "I told you so" with the benefit of hindsight.[57] On the other hand, Macrides highlights Psellos's inclination (as vain and self-centred as he was) to use this literary tool to draw attention to *his* role in making history and to lure his audience to sharing in the process, thus making himself "the subject to which one should always return."[58] Greatly influenced by Psellos, Anna also uses her interjections, according to Macrides, to draw the audience's attention to herself as a woman and a writer, her self-consciousness, and her confidence in handling her material in such a way that she would not be smeared as a panegyrist of her family.[59]

Ethnographic Accounts of Foreign Peoples

A third requirement means that the narrative of Byzantine historians would include ethnographic accounts of foreign peoples. As the Byzantines confronted numerous nations in different operational theatres of war during their long history, the genre of Byzantine ethnography that has received great attention is the military manuals, also known as *strategika* or *taktika*, which can be defined as a specific category of literary works devoted to war.[60] These ethnographical surveys of foreign peoples that are found

---

55 J. N. Ljubarskij, "'Writers' Intrusion" in Early Byzantine Literature," *XVIIIth International Congress of Byzantine Studies: Major Papers* (Moscow, 1991), 433–56, especially 435.

56 Ruth Macrides, "The Historian in the History," in *Philellen: Studies in Honour of Robert Browning*, ed. Costas N. Constantinides, Nikolaos M. Panagiotakes, Elizabeth Jeffreys, and Athanasios D. Angelou (Venice: Biblioteca dell'Istituto ellenico di studi bizantini e postbizantini di Venezia, 1996), 205–24 (quotation at 209).

57 Attaleiates, *History*, 232–41; Bryennios, *Ὕλη Ἱστορίας*, 79–81; Macrides, "The Historian in the History," 209–10.

58 Macrides, "The Historian in the History," 215–16.

59 Macrides, "The Historian in the History," 217–20; Neville, *Anna Komnene*, 75–88; Jessica Doyle, "Anna Komnene's Use of Personal Intrusion in the *Alexiad*," (MA thesis, Ohio State University, 2012), especially chapters 3 and 5.

60 For a more detailed definition of this literary genre: Georgios Theotokis, "Some Timeless

in these *taktika* showcase two critical aspects of the Byzantine "military culture,"[61] their ability to scrutinize and evaluate the tactics and fighting abilities of their enemies, and their eagerness to learn from them when necessary and adapt their tactics to the requirements of each operational theatre.

The latter point is perhaps best illustrated in the *Sylloge Taktikorum*'s (ca. 930) writing that "Most of the races of the various peoples [fighting the Byzantines] mount ambuscades, especially the Persians, Turks, and Arabs. But sometimes the Romans mount ambuscades as well, having in fact been taught by these races. For after the Romans had suffered many times, they learned by experience from heavy losses and they contrive against this very thing."[62] This idea of learning from and adapting to the tactics of their enemy was deeply embedded in the Byzantine psyche, as the evidence from the *taktika* and several historical sources clearly show.[63]

Regarding the former aspect, the author of the *Strategikon of Maurice* (ca. 600) emphasizes that "that general is wise who before entering into war carefully studies (*πολυπραγμονῶ*) the enemy."[64] He then links the emphasis placed (in Book VII) on the gathering of intelligence about the enemy with his famous ethnographic study in Book XI ("Characteristics and Tactics of Various Peoples"), where he reiterates that "the purpose of this [XI] chapter is to enable those who intend to wage war against these peoples to prepare themselves properly [*ἱκανῶς ἁρμόζεσθαι*]."[65] In fact, the *Strategikon* was the first to devote a substantial section of the work on the polity and the social and ethnic divisions of all the nations that bordered the empire, their geographical and ethnological backgrounds, their political, social and military organization, and the stereotypical characteristics/clichés about their fighting prowess that we can also see in Sextus Julius Frontinus's *Strategemata* and Flavius Vegetius Renatus's *Epitoma Rei Militaris*.[66]

Traits of War and the Vertical Transmission of Military Knowledge through the *Taktika*," in *Military Manuals in the Medieval Mediterranean and Beyond*, ed. Conor Whately (Leiden: Brill, forthcoming).

61 "'Military social structure'can be defined as the structure that consists of the arrangement into military groups, such as divisions, battalions, and companies; 'military culture,' conversely, can be understood as the 'operational code of war" that is followed by an entire nation or people or, to give a more straightforward definition, it could mean "a way of understanding why an army acts as it does in war." I. V. Hull, *Absolute Destruction, Military Culture and the Practices of War in Imperial Germany* (Ithaca: Cornell University Press, 2005), 93–94; J. L. Soeters, D. J. Winslow, and A. Wibull, "Military Culture," in *Handbook of the Sociology of the Military*, ed. G. Caforio (New York: Springer, 2003), 237–54. For an analysis of the terms "military culture" and "militarism" and their distinctive features: P. H. Wilson, "Defining Military Culture," *Journal of Military History* 72 (2008): 11–41.

62 Georgios Chatzelis and Jonathan Harris trans., *A Tenth-Century Byzantine Military Manual: The Sylloge Tacticorum* (London: Routledge, 2017), 24.1.

63 On the diffusion of military knowledge between the Byzantines and the Arabs in Cilicia, Syria, and northern Mesopotamia in the mid-tenth century: Theotokis, *Byzantine Military Tactics*.

64 Maurice's *Strategikon*, 7.

65 Maurice's *Strategikon*, 11. Compare with: *The Taktika of Leo VI*, cons. 12, para. 106, 181; Kekaumenos, Στρατηγικόν, 52–54.

66 Frontinus, *Strategemata*, 2.3.1, 2.5.31; Vegetius, *Epitome*, 1. 1–2; Machiavelli, *Art of War*, 1. 121. On the perception of the "other" in Byzantine "military" sources: Różycki, *Battlefield Emotions*,

A critical aspect of the *Strategikon*'s analysis is the author's repeat of negative contrasts to stress the difference between the Romans and their enemies' tactics—they do not attack like X (like the Romans do), but instead, they attack like Y (a unique or peculiar tactic).[67] Moreover, the authors of the *Strategikon* and the *Taktika of Emperor Leo VI* (ca. 904–916), the latter deriving its basic format and inspiration from the former,[68] were both deeply interested in their enemies' *κρᾶσις*—the physical strength and endurance of warriors in adverse conditions. They built their discussion around the offensive and defensive equipment of the foreign nations surrounding the empire, their camps and battle formations, followed by the identification of weaknesses that each nation eventually displays on the battlefield, and advice on how to take full advantage of them.[69] The climatic clichés remained the same in both works, despite the three centuries that set them apart; for example, Leo's views on the Arabs as a people who prefer heat and dislike cold and rain because it slackened their bowstrings to echo Maurice's comments on the Persians! Leo copied this cliché from the *Strategikon*, although he failed to inform the reader about the "classified" material that apparently left the Byzantine bowstrings intact in similar weather conditions![70] We need to point out, however, that these literary *topoi* are missing from the mid-tenth-century *taktika* commissioned by Nikephoros Phokas, the "On Skirmishing" and the *Praecepta Militaria*, which depict a more "sober" portrait of the Arab enemy that was painted by military professionals whose battlefield experience in Syria and Mesopotamia span many decades and had filled them with great respect for their Arab adversaries.[71]

The great value of these manuals rests largely on their quality as repositories of (military) knowledge and on their didactic use for army officers and civil servants.[72] Their particular purpose as utilitarian texts that commanders used in preparation for campaigns or combat has been frequently emphasized; for example, Walter E. Kaegi, Jr. has argued that it was the *Strategikon of Maurice* that would have proved most useful to Heraclius and his advisers in the military planning against the Persians, much more

85–93; Theotokis, *Byzantine Military Tactics*, chapter 4; Kaldellis, *Ethnography after Antiquity*, 82–87. For the historical evidence that can be deducted from the *Strategikon* on the military tactics and organization of the empire's neighbouring éthnē: Łukasz Różycki, "The *Strategikon* as a Source: Slavs and Avars in the Eyes of Pseudo-Maurice, Current State of Research and Future Research Perspectives," *Acta Archaelogica Carpathica* 52 (2017): 109–31; Georgios Kardaras, "Το σχήμα των Αβάρων στο Στρατηγικόν του Μαυρικίου: Μια κριτική προσέγγιση," *Byzantinos Domos* 16 (2007–8): 151–67; John Earl Wiita, "The Ethnika in Byzantine Military Treatises" (PhD diss., University of Minnesota, 1977).

67 Maurice's *Strategikon*, bk. 11, 1.

68 John Haldon, *A Critical Commentary on the Taktika of Leo VI* (Washington DC: Dumbarton Oaks, 2014), 331–88.

69 Detailed discussion in Theotokis, *Byzantine Military Tactics*, 106–12.

70 Riedel, *Leo VI*, 45–46.

71 Theotokis, *Byzantine Military Tactics*, 118–20.

72 Conor Whately, "The Genre and Purpose of Military Manuals in Late Antiquity," in *Shifting Genres in Late Antiquity*, ed. G. Greatrex and H. Elton (Farnham: Ashgate, 2015), 255–61.

than the works of Procopius, Agathias, and Menander.[73] On the other hand, Whately more recently suggested that the *Strategikon* and Procopius's *Wars* more or less complemented each other in the eyes of their audience and should both be studied in parallel for their advisory nature on how to defeat the Persians, Vandals, Goths, and like-minded peoples in combat, because Procopius's experience and expertise put him in an ideal position to discuss these matters and his *Wars* include several features unique to particular operational theatres.[74] Yet, as Kaldellis points out, the historiographical model of Procopius and Agathias was not followed by the historians of the Middle Byzantine period, positing a major "rupture" in the seventh century, after which there was a noticeable decline in (military) ethnography.[75]

Nonetheless, ethnography does not disappear after the seventh century and modern historians can still encounter several ethnographical accounts embedded in histories of the Middle Byzantine period that are rather brief or partial in their coverage compared to their late Antique or classical predecessors. This is usually the case when a new geopolitical player comes into view and infiltrates the Byzantine Empire's sphere of influence, at which point the author felt obliged to offer a brief digression into the historical and ethnographical background of this "newcomer," one that also explained the geopolitical movements that led them to their current location. Hence, if we focus on the great geopolitical movements that troubled the empire in the middle of the eleventh century, namely the Seljuk and Patzinak invasions of Anatolia and the Balkans, we may ask what sort of information can be extracted from our primary sources about the Byzantine perception of the "military other," first regarding their competence and skills as warriors and, second, any social clichés or literary *topoi* that can be found in the narrative.

Kaldellis calls the ethnographic digressions on the origins of the Seljuks and the Patzinaks in the works of Skylitzes, Attaleiates, Bryennios, and Komnena as a "quasi-ethnographic" subgenre of Byzantine historiography that simply tracks down group movements into different geographical locations, rather than offering an *origo gentis* of the people in the classical model.[76] And that is true in the sense that none of the aforemen-

---

73 Walter E. Kaegi Jr., *Heraclius: Emperor of Byzantium* (Cambridge University Press: Cambridge, 2003), 74.

74 Whately, *Combat, Culture, and Didacticism*, 4–5, 228–30.

75 Kaldellis, *Ethnography After Antiquity*, chapter 3.

76 Kaldellis, *Ethnography After Antiquity*, 98 (analysis at 93–98). See also: Hans Werner Goetz, Jörg Jarnut, and Walter Pohl (eds.), *Regna and Gentes: The Relationship Between Late Antique and Early Medieval Peoples and Kingdoms in the Transformation of the Roman World* (Leiden: Brill, 2003). Some interesting parallels can be drawn with the "construction" of the Norman "myth" of invincibility on the battlefield: Matthew Bennett, "Norman Conquests: Nature, Nurture, Normanitas," in *The Normans in the Mediterranean*, ed. Emily A. Winkler and Liam Fitzgerald (Brepols: Turnhout, 2021), 43–66; Emily A. Winkler and Liam Fitzgerald, "Norman Battle Tactics in the Mediterranean Theatre of Operations: Fighting Lombards, Greeks, Arabs, and Turks c.1050–c.1100," in *Warfare in the Norman Mediterranean*, ed. Georgios Theotokis (Woodbridge: Boydell , 2020), 133–50; Francesca Petrizzo, "'Conquest in Their Blood': Hauteville Ambition, Authorial Spin, and Interpretative Challenges in the Narrative Sources," in Theotokis, ed., *Warfare in the Norman Mediterranean*, 35–54.

tioned authors reveals anything about these "newcomers" other than their geographical origins, their religious beliefs, and the actions of their leaders that brought them into Byzantium's sphere of influence (or, "too close to be ignored"); nothing about the culture, religious practices, or socio-political organization of these peoples is considered in any detail.[77] But what about their skills as warriors?

Byzantine historians disappoint modern scholars with the information they provide about the political and military organization, and fighting tactics of the Patzinaks, and the amount of information only becomes tantalizingly more detailed when these people caused political and military unrest.[78] Already for a century before their (in)famous crossing of the Danube in 1046/47, the Patzinaks had earned a fierce reputation in Byzantine sources,[79] with the *De Administrando Imperio* (ca. 948–952) painting a picture of the 'scythian" people as foreign/alien to the empire,[80] echoing the literary *topoi* that described the "people of the North" as independent and numerous,[81] but untrustworthy, insatiable (*ἄπληστοι*), and shameless in their demands for generous gifts,[82] and in general as the "devil's brats."[83] Attaleiates, Psellos, and (much less) Skylitzes paint a similar picture of the Patzinaks as an alien but independent and numerous nation that was "not used to fighting against the Roman phalanxes," instead showing preference to "armed raids more than any other skill or art," echoing Book XI of the *Strategikon of Maurice* on the "Avars, Turks, and Others Whose Way of Life Resembles That of the Hunnish Peoples."[84]

Although it must be said that the image of the Patzinaks in the Byzantine sources is not always homogenous,[85] they are portrayed as a terrifying people on the move,[86] who

---

77 John Skylitzes, *A Synopsis of Byzantine History*, 811–1057, trans. John Wortley (Cambridge: Cambridge University Press, 2010), 416–20 (on the Seljuks) and 426–31 (on the Patzinaks); Attaleiates, *History*, 77–81; Bryennios, *Ὕλη Ιστορίας*, 61–73, Bryennios reproduced the digression on the Turks in the *Synopsis* almost word for word in his Hyle Historias: Wortley, *A Synopsis of Byzantine History*, xxxii.

78 Marek Meško, "Pecheneg groups in the Balkans (ca. 1053–1091) According to the Byzantine Sources," in *The Steppe Lands and the World Beyond Them: Studies in Honor of Victor Spinei on his 70th Birthday*, ed. Florin Curta and Bogdan-Petru Maleon (Iasi: Editura Universitii "Al. I. Cuza," 2013), 179–205.

79 Elisabeth Malamut, "L'image byzantine des Petchénègues," *Byzantinische Zeitschrift* 88 (1995): 105–47 at 112–21.

80 *De Administrando Imperio*, 67 (§13). Marek Meško, "Nomads and Byzantium: Problematic Aspects of Maintaining Diplomatic Relations with the Pechenegs," in *On Research Methodology in Ancient and Byzantine History*, ed. Jarmila Bednarikova, Marek Meško and Anna Zakowa (Brno: Tribune, 2015), 181–93.

81 *De Administrando Imperio*, 57 (§8); Maurice's *Strategikon*, bk. 11, 2.

82 *De Administrando Imperio*, 55 (§7, §8).

83 *De Administrando Imperio*, 57 (§8).

84 Attaleiates, *History*, 53–59; Psellos, *Chronographia*, (ed. Sathas), [7:68] 222; Maurice's *Strategikon*, bk. 11, 2; Skylitzes, *Synopsis*, 426; Malamut, "L'image byzantine," 122.

85 Malamut, "L'image byzantine," 126.

86 Anna Komnena's account of the Battle of Levounion (1091) shows that the Patzinaks still (forty

instilled fear in the hearts of their enemies,[87] who do not fight like the Romans and enjoy pillaging the lands of the Balkans. There are some glimpses of their fighting tactics in the sources, including those employed by the Byzantines as a countermeasure, yet these are just glimpses. Attaleiates notes that when the commander of the Western armies, the *magistros* Konstantinos Arianites, marched against the Patzinaks from Adrianople, he suffered a heavy defeat near Diampolis (June 1050) because his troops "recklessly attacked them [Patzinaks] before the city in cultivated fields broken up by vineyards and fences," thus handing over the tactical advantage to the more numerous and mobile nomads instead of "guarding themselves inside their camp...and pursuing the Skythians closely from behind when they were on their return journey," a tactic recommended in the mid-tenth-century military treatise known as "On Skirmishing."[88] A few years later, the Byzantines did allow the Patzinak raiders to disperse into smaller parties before their troops, who were sheltering in fortified towns, rushed out and attacked them while the raiders were disorganized and laden with loot, another tactic that was favoured by the author of the "On Skirmishing."[89] The fighting tactics of the Patzinaks become somewhat clearer when Attaleiates describes an episode of them falling upon an army commanded by Botaneiates, with the nomads riding "in circles around the group [of Byzantines] on their horses and fired swarms of arrows into it"; Botaneiates's reaction involved forming into a compact unit to protect against the "arrow-showers' and to redeploy his men on a river bank to avoid a possible encirclement by the enemy-sound advice that can also be found in the *Strategikon*.[90] Forming a compact formation to protect against the Patzinak arrows will showcase Alexios Komnenos's skills as a tactician at the Battle of Dristra (August 1087), when he ordered his troops not to break ranks until the Patzinaks were close enough—"no more than a bridle's length."[91] In April 1091, Alexios deployed his troops at Levounion in a crescent-shaped (μηνοειδής) formation to counter the well-known encircling manoeuvres of the nomads.[92]

Ethnographic images of the "Turks" (i.e., the people known as Magyars or Hungarians) in Leo VI's *Taktika* and the *De Administrando Imperio* again reflect earlier information and stereotypes drawn from the *Strategikon of Maurice*.[93] Like the "Scythians,"

---

years after their crossing of the Danube) were a nation on the move, with women, children, cattle, horses, and all movable property in train. Komnena, *Alexiad*, 8:5.

87 "It is said that they [Romans] were put to flight before they even heard the hoof-beats of the [Patzinak] horses." Skylitzes, *Synopsis*, 438.

88 Attaleiates, *History*, 59; "On Skirmishing," Dennis (ed.), *Three Byzantine Military Treatises*, §§3–4.

89 Attaleiates, *History*, 63; "On Skirmishing," Dennis (ed.), *Three Byzantine Military Treatises*, §§8–9.

90 Attaleiates, *History*, 71; Maurice's *Strategikon*, bk. 11, 2; bk. 12, 22.

91 Komnena, *Alexiad*, bk. 7, 3.

92 Komnena, *Alexiad*, bk. 8, 5.

93 Haldon, *A Critical Commentary*, 338–47; Jonathan Shepard, "Byzantine Writers on the Hungarians in the Ninth and Tenth Centuries," in *Emergent Elites and Byzantium*, ed. Jonathan Shepard (Aldershot: Variorum, 2011), 97–119.

the "Turks" coming from the North are depicted as undisciplined[94] (as opposed to the Romans and the Persians/Arabs who were their sedentary neighbours), very numerous[95] (again, in contrast to their sedentary neighbours), proud, and independent, but they are also portrayed as being deceitful "by nature" (*Taktika*: κρύπτοντα την βουλήν ἀυτῶν; *Strategikon*: κρυψίβουλα), untrustworthy (even duplicitous) in their agreements (ὅρκου περιφρονοῦσι), and insatiable (ἄπληστοι) in their demands for gifts.[96] Leo's account of the "Turks," as expected, is replete with ethnographical *topoi* about the nomads from the steppes and their mastery of the horse and the bow in warfare, including their employment of encircling manoeuvres and feigned retreat tactics in battle.[97]

Unfortunately, Byzantine historians are equally disappointing as they are about the Patzinaks when it comes to providing information about the military organization and fighting tactics of the Seljuk Turks in the eleventh century. The first battle between the imperial forces and the Seljuks that drew the attention of the Byzantine sources in more detail is the Battle of Kapetron[98] (modern Hasankale, Erzurum province; September 19, 1048), where the *doux* of Ani Katakalon Kekaumenos, the *catepan* of Vaspourakan Radomir Aaron, and the imperial *magistros* Liparit IV of Kldekari clashed with the Turkish forces of Ibrāhim Yināl some forty kilometres east of Theodosiopolis. Although we know more about the strategic movements of both forces in the days leading up to the battle, including Skylitzes's narrative about the Byzantine commanders' inability to counter the Seljuk invasion with an effective strategy, we learn frustratingly little about the battle itself. Aristakēs Lastivertc'i, Matthew of Edessa, and Attaleiates only emphasize the capture of Liparit by the Turks,[99] while Skylitzes considers the breaking of the Turkish flanks by the divisions led by Kekaumenos (right wing) and Aaron (left wing) as the turning point of the battle without going into any details.[100]

Our sources are brief and vague in their description of battle tactics for Tughril Beg's second large-scale campaign in the summer of 1054 that targeted the key fortresses of Perkri (modern Muradiye), Artchesh (modern Erciş) and Manzikert around Lake Van. However, we get a tantalizing glimpse of a clash between Turkish and some Varangian troops that were defending the city of Paipert (modern Bayburt) in the theme of Chaldea, some 140 km northwest of Theodosiopolis; we read in Aristakēs's account:

---

94 *The Taktika of Leo VI*, cons. 14, 38; cons. 18, 38.

95 *The Taktika of Leo VI*, cons. 18, 43; Maurice's *Strategikon*, bk. 11, 2.

96 *The Taktika of Leo VI*, cons. 18, 45; *De Administrando Imperio*, 67 (§13). The desertion of Patzinak allies of Romanos IV at Manzikert confirms the nomadic topos of fickleness and untrustworthiness: Leo VI, *Taktika*, bk. 18, 64. Compare with Aristakēs Lastivertc'i's portrayal of the Seljuk Turks as falling upon the Anatolian Christians like "ravenous wolves." Aristakes Lastivertc'i's *History*, 64, 108, 123, 143.

97 *The Taktika of Leo VI*, cons. 18, 47–62.

98 For the campaign leading up to and for the Battle of Kapetron: Leveniotes, *Η πολιτική κατάρρευση του Βυζαντίου*, 92–97.

99 Aristakes Lastivertc'I, *History*, 88; Matthew of Edessa, *Chronicle*, 79; Attaleiates, *History*, 79.

100 Skylitzes, *Synopsis*, 425.

> There they [Turks] encountered a brigade of Byzantine troops called Vrhangs (Vrangk') who, at all hazards, battled with them. By God's mercy, the Byzantine brigade grew stronger, vanquished the enemy, killed the head of their troops and many with him, turned the rest to flight and retrieved all the loot and slaves. However, they did not dare to pursue the fugitives very far, since they were afraid of encountering a large force.[101]

The clash of heavily armed and disciplined Varangian infantry with Turkish nomads renowned for their mobility and use of encircling tactics would have offered a great opportunity to study the cultural reaction of both parties in what, certainly, was a transcultural conflict in eastern Anatolia, but Aristakēs's account does not give us many details with which to work. Nor does Matthew of Edessa's description tell us how the Armenians from Sasun (in Taron) fell upon a Turkish raiding party, in 1058, that resulted in the routing of the enemy's left flank.[102] Another transcultural clash about which it would be intriguing to know more is the one between the troops led by the Frankish commander Hervé Frangopoulos who, after having broken with the emperor, allied with a Turkish chieftain named Samouch, who had his base in Vaspourakan. If we believe the sources, however, this alliance did not last for long: discord quickly developed into a battle (in 1059?) in which many Turks were killed, yet the only details we have for the battle is a brief reference to the cunning nature of the Turks who "attacked the Franks as they were dining."[103] The ethnographical *topoi* about the Turks and their mastery of the horse and the bow in warfare, including their feigned retreat tactic which, by Manuel Komnenos's summer campaign of 1070, Attaleiates would describe as "one of their usual tricks and stratagems," these become somewhat more numerous in our sources the closer we get to the fateful campaign of 1071. I will examine these *topoi* later as part of my analysis of the battles for the campaigning seasons between 1068 and 1071.

A Narrative Rich in Battles and Sieges often Preceded by Speeches

To be regarded as authoritative as that of their classical predecessors, medieval historians had to meet a fourth requirement; their narrative had to be rich in battles and sieges, often preceded by speeches and even letters. McGrath distinguishes between two types of information that are included in the "battle-piece";[104] the first type would be the essential facts regarding the event—when and where the battle took place and what description we have of the battlefields and army campaign routes, what was the size and composition of the opposing armies, and what record there is of who won and who lost the battle, including battle manoeuvres and the tactics employed by the opposing generals. I will examine the accounts of the historians of the Battle of Manzikert in terms

101 Aristakes Lastivertc'I, *History*, 97.

102 Matthew of Edessa, *Chronicle*, 93.

103 Skylitzes, *Synopsis*, 453; Matthew of Edessa, *Chronicle*, 99–100.

104 McGrath, "Warfare as Literary Narrative," 169.

of the figures they provide, and their topographical knowledge of the campaigns of the years between 1068 and 1071, in the next chapter, whilst the following section of this chapter will focus on the details and accuracy of their description of the clashes between the opposing armies. These basic facts formed the skeletal structure of the "battle-piece" that most historians embellished with a second layer of detail; this included the exhortative speeches delivered before or during battle, prayers and private or public religious rituals meant to secure the aid of the divine,[105] and descriptions of heroic episodes from individual or group champions.[106]

Exhortation speeches had already become a traditional feature of high historiography in classical Greece and Rome,[107] while Muslim chroniclers were also well aware of the dramatic value of inserting speeches into their work at moments of emotional intensity.[108] The military treatises of Late Antique Rome and Byzantium considered the primary purpose of the exhortation speech was to raise and bolster morale,[109] for classicizing historians like Polybius, Procopius, Leo the Deacon, and Attaleiates (among many others), who had Thucydides as a model of military oratory, speeches served as an effective "narrative tool" for historical episodes, to help the audience better understand and evaluate the respective leaders' performances and assess the real causes of a victory or defeat.[110] Furthermore, the insertion of battle exhortations into the narrative, regardless if the speech that has survived was a trustworthy reproduction of the actual speech

---

105 Konstantina Karaple, *Κατευόδωσις στρατού: Η οργάνωση και η ψυχολογική προετοιμασία του Βυζαντινού στρατού πριν από τον πόλεμο (610–1081)* [Kateuodosis Stratou: The Organization and Mental Preparation of the Byzantine Army Before War (610–1081)] (Athens: Myrmidones, 2010), 85–89.

106 Savvas Kyriakides, "Accounts of Single Combat in Byzantine Historiography: 10th–14th Centuries," *Acta classica: Verhandelinge van die Klassieke Vereniging van Suid-Afrika* 59 (2016), 114–136; Theocharis Alexopoulos, "Η προσωπική μονομαχία στον Βυζαντινό κόσμο (11ος–15ος αι.)" [Personal Duel in the Byzantine World (Eleventh–Fifteenth Centuries)] (PhD diss., University of Athens, 2016); Triantafyllitsa Maniati-Kokkini, "Η επίδειξη ανδρείας στον πόλεμο κατά τους ιστορικούς του 11ου και 12ου αι" [Demostrating Bravery in War According to the Historians of the Eleventh and Twelfth Centuries], in *Byzantium at War, 9th–12th c.*, ed. by K. Tsiknakes (Athens: National Hellenic Research Foundation, 1997), 239–59.

107 Theotokis, Georgios, and Dimitrios Sidiropoulos, *Byzantine Military Rhetoric in the Ninth Century: A Translation of the Anonymi Byzantini Rhetorica Militaris* (London: Routledge, 2021), 28–35; John Marincola, "Speeches in Classical Historiography," in *A Companion to Greek and Roman Historiography*, ed. by John Marincola (Oxford: Blackwell, 2007), 1:118–32.

108 See the discussion about the Futūḥ al-buldān as conquest literature in Arabic historiography: Ryan J. Lynch, Arab Conquests and Early *Islamic Historiography*, The Futuh al-Buldan of al-Baladhuri (London: Bloomsbury, 2020), 156–60; Tarif Khalidi, Arabic Historical Thought in the Classical Period (Cambridge: Cambridge University Press, 1994), 65–68.

109 Theotokis and Sidiropoulos, *Byzantine Military Rhetoric*, 35–54.

110 Polybius, Histories, 12.25; Whately, *Combat, Culture, and Didacticism*, 77–84, 134–39; Treadgold, *Middle Byzantine Historians*, 244; Kaldellis, *Procopius of Caesarea*, 29–34. Compare with: Anthony Kaldellis, "The Original Source for Tzimiskes' Balkan Campaign (971 AD) and the Emperor's Classicizing Propaganda," *Byzantine and Modern Greek Studies* 37 (2013): 35–52; Anna Kotlowska and Łukasz Rozycki, "The Role and Place of Speeches in the Work of Theophylact Simocatta," *Vox Patrum* 36 (2016): 353–82; Joseph D. C. Frendo, "History and Panegyric in the

delivered to the troops or if it was a literary invention,[111] undoubtedly contributed to the theatricality and liveliness of the "battle-piece," while adding moralistic/didactic comments regarding the role of providence and fortune that the audience could pick up and "decipher."[112]

Speeches in Attaleiates, Skylitzes, and Bryennios are few and usually short. In fact, the former two refer to the emperor Romanos as "assembling the army [to give] a speech about the coming battle [on August, 26]," yet both of them do not go into any details about the actual speech that would have been delivered to the troops. However, the terms *τραχυτέρων λόγων/ρημάτων*[113] points to the category referred to by Syrianos Magistros as "the reproach speech [that] is to some extent inspired by the consolatory, which is why it is closely related to it, as both of them urge a future display of courage. For you [general] must not only rebuke those who have been defeated because of sluggishness in battle, but at the same time comfort them, and through consolation and exhortation motivate them to war."[114] Indeed, both sources mention the emperor's speech after the initial setbacks between Bryennios "the Elder" and Basilakes's units and the Seljuks; interestingly, the author of the *Hyle* muffles Romanos's speech, mentioning, instead, an exhortation speech that was delivered to his soldiers by Bryennios "the Elder," calling on them to prove themselves worthy of the bravery of their Roman ancestors.[115]

Alp Arslan's speech by al-Jawzī on the eve of the Battle of Manzikert is, perhaps, one of the most impassionate and evocative speeches put in the mouth of a Muslim leader for many centuries:

> We are with a depleted number of men. I want to throw myself on them [the Byzantines] at this hour when prayers are being said for us and for the Muslims on the pulpits. Either I will achieve the goal or I will go as a martyr to Paradise.

Age of Heraclius: The Literary Background to the Composition of the 'Histories' of Theophylact Simocatta," *Dumbarton Oaks Papers* 42 (1988): 143–56.

111 It is fair to say that what modern historians get to read is—in all probability—a reconstructed and literary elaborated version of a much shorter speech that would have actually been circulated on the battlefield: Ralph-Johannes Lilie, "Reality and Invention: Reflections on Byzantine Historiography," *Dumbarton Oaks Papers* 68 (2014): 157–210; Edward Anson,"The General's Pre-Battle Exhortation in Graeco-Roman Warfare," *Greece & Rome* 57 (2010): 304–18; Mogens Herman Hansen, "The Little Grey Horse: Henry V's Speech at Agincourt and the Battle Exhortation in Ancient Historiography," *Histos* 2 (1998): 46–63; W. K. Pritchett, "The General's Exhortation in Greek Warfare," in *Essays in Greek History* (Amsterdam: Gieben, 1994), 27–109; Mogens Herman Hansen, "The Battle Exhortation in Ancient Historiography: Fact or Fiction?," *Historia: Zeitschrift für Alte Geschichte* 42 (1993): 161–80.

112 McGrath, "Warfare as Literary Narrative," 173.

113 Attaleiates, History, 280–81; Skylitzes, *Continuation*, 116–17.

114 Theotokis and Sidiropoulos, *Byzantine Military Rhetoric*, 87.

115 Bryennios, *Ύλη Ιστορίας*, 86–87. The general's aim was simply to remind them of their responsibility in emulating the achievements of their ancestors: Theotokis and Sidiropoulos, *Byzantine Military Rhetoric*, 40.

> He amongst you who wants to follow me, let him follow me, and he who wants to leave, let him leave my company. Here is not a sultan commanding, nor an army being commanded, for today I am only one of you and a ghāzī with you. He who follows me and gives himself to God Most High, he will gain Paradise and booty. He who goes away, the Fire and ignominy are obligatory for him.
>
> They replied: O Sultan! We are your servants and whatever you do we will follow you in it and help you in it. So do what you want.[116]

The idea of triumphing against all (numerical) odds will be analyzed in more detail in the following chapter, where I will compare Alp Arslan's victory to Muhammad's triumph at Badr, and with the construction of an image of the Normans as "model warriors" who overpowered their more numerous enemies on the battlefields of Italy and Sicily; the emphasis there will be the victory of the underdog over a (numerically) superior adversary–the 'superpower" of the age. Here, the emphasis is on the "moral struggle" between two leaders that will assign history with a hero and an "anti-hero."

First, we need to remind ourselves that the author of this speech was writing his universal chronicle in Baghdad, in the final decades of the twelfth century, a period of intense activity against the Crusaders, when Muslim chroniclers were showing a lively interest in the history of the Great Seljuks, and in particular the Battle of Manzikert. In his effort to exalt Alp Arslan, al-Jawzī assigns to him several Quranic virtues that aim to portray him as the "model jihad warrior" of the age, and where the hero of the campaign—the sultan—is pitted against an anti-hero—the emperor—in the mindset of his reader(s), who should have been able to draw their conclusions about the righteousness of the Seljuk cause. Hence, Alp Arslan demonstrates courage and mental strength that gives rise to a military prowess based on the firm belief in God and the Day of Judgement. For that reason we should not doubt that he was dressed in white that fateful day, as some Muslim accounts maintain; the true believer is 'God-fearing' but does not show fear against his human enemy.[117] The sultan demonstrates, but also demands of his men, one of the fundamental social virtues in a nomadic world—loyalty, promising to those who follow him in 'true heart' rewards both in this world and in the next, and this virtue is one that only a true leader could ask.[118] In fact, loyalty and obedience being the most tangible demonstration of loyalty, put the believers in sharp contrast to that of unbelievers. Likewise, the generous, brave, and modest attitude of believers is very different from the arrogant, ruthless, and selfish behaviours of unbelievers.[119] Hence the dramatic contrast between Romanos's arrogant stance that precedes Alp Arslan's speech, where al-Jawzī emphasizes not only the extravagance of the Byzantine army's baggage train, with the ridiculously high numbers of men and material, but also highlights

116 al-Jawzī, in: Hillenbrand, *Turkish Myth*, 38–39.

117 Toshihiko Izutsu, *Ethico-Religious Concepts in the Qur'an* (Montreal: McGill–Queen's University Press, 2002), 83–86.

118 Izutsu, *Ethico-Religious Concepts*, 86–97.

119 Izutsu, *Ethico-Religious Concepts*, 119–55.

the emperor's arrogance, insolence, and vengeful attitude in rebuffing the sultan's overtures of peace, thus clearly painting an anti-hero: "I have spent a lot of money and assembled many troops to come to the like of this situation. If I am victorious in it, how could I leave it? How preposterous! There will be no treaty except in Rayy and no going home except after I have done in the lands of Islam the like of what has been done in the lands of Byzantium."[120] Therefore, the effect of Romanos's arrogant stance further enhanced in the Muslim reader's mind the triumph achieved by the humble and weakened Seljuk army. For al-Jawzī, it was "mission accomplished."

To return to the point made earlier, in most battle descriptions, the basic presentation of facts is brief and the second category of detail, the authorial embellishment, constitutes most of the battle episode. Shaw, in his comprehensive study on war and violence in Late Antiquity, referred to the "daunting problems [that] are posed by the written sources" for the "reasoned understanding of war," to uphold a bit further down that "classical historiography and allied forms of literature had developed canons of reporting violence that were stereotypical and schematic, and that depended on typecast descriptions derived from literary prototypes. These conventions compelled the writer to respect the limits of what was acceptable in the description of violence, even if he was a historian who had been an eyewitness to the events."[121] But should we rush to dismiss the late antique and medieval battle accounts as derivative and unoriginal? It would be unhelpful, and, instead, we need to bear the author's socio-economic and cultural background in mind.

The Artificiality of the "Battle-piece": The Role of the Author's Cultural Milieu

Many theoretical discussions of ancient and medieval historiography inevitably turn to the role of two factors that contributed to the artificiality of the descriptions of battles: culture and rhetorical education. In 1999, Lendon argued that "battle descriptions are products as much of culture as of observation," taking his cue from the study of Caesar's (understanding of) battle description that was, according to the author, "further from our own than are those of his [Caesar's] predecessors," thus leading him to suggest that "the way ancient authors describe the details of battle can tell us about the mental rigging of the societies in which they lived."[122] Before Lendon, McGrath had also shown that any differences in the battle descriptions in the *Synopsis* of John Skylitzes and the *History of Leo the Deacon* should be attributed to the contrasting values, level of understanding (experience), and editorial

120 al-Jawzī, in Hillenbrand, *Turkish Myth*, 38.

121 Brent D. Shaw, "War and Violence," in *Late Antiquity: A Guide to the Postclassical World*, ed. G. Bowersock, P. Brown, and O. Grabar (Cambridge, MA: Belknap, 1999), 130–33.

122 J. E. Lendon, "The Rhetoric of Combat: Greek Military Theory and Roman Culture in Julius Caesar's Battle Descriptions," *Classical Antiquity* 18 (1999): 273–329 (quotations from 274–75).

decisions (personal biases, political "agendas," etc.) of the two historians.[123] More recently, in his meticulous study of Procopius's "battle-pieces," Whately maintained that "the key to understanding Procopius's descriptions of combat is to pay attention to his didactic aims, which were determined by the conventions of classicising history and the interests of his audience."[124] According to Whately, the author's cultural background (the values, norms, and assumptions that guide human action) and his understanding of the "military culture"—the "operational code of war" that is followed by an entire nation or people are clearly reflected in Procopius's descriptions of combat.[125]

Finally, the body of chronicles of the First Crusade is especially well-suited for discussing the problem of oral sources and their records/descriptions of historical episodes (including battles and sieges) in medieval writings of history. Harari analyzed the passages in which chroniclers who participated in the First Crusade claimed to have been present at certain points in the expedition.[126] He highlighted that in order for modern historians to be able to examine the authorial intention of their sources, they have to be mindful of how they treat "first-hand accounts" in relation to what kind of text they intend to produce. Critically, they have "to investigate what contemporaries of the crusade understood by 'facts,' what kinds of facts they deemed important, and what kind of importance these facts were accorded."[127] Picking up from Harari's point, Lapina showed that records can be strongly influenced by the ideals of the authors and their subsequent experiences, and the criteria by which these authors considered evidence were defined to a great extent by the intellectual and theological requirements of the twelfth century.[128] Lapina placed particular emphasis on the theological notion of "witnesses" and the significant impact it had on historical writing in the Middle Ages, where second-hand chroniclers scrutinized and adapted historical episodes (particularly miracles, but also battles and sieges) to a more conventional historical narrative model because they believed they understood the historical and theological significance of those events better than participants had.

---

123 Stamatina McGrath, "The Battles of Dorostolon (971), Rhetoric and Reality," in *Peace and War in Byzantium*, ed. T. S. Miller and J. Nesbitt (Washington, DC: Catholic University of America Press, 1995), 152–64.

124 Whately, *Combat, Culture, and Didacticism*, 4.

125 Whately, *Combat, Culture, and Didacticism*, 6–7.

126 Yuval Noah Harari, "Eyewitnessing in Accounts of the First Crusade: The Gesta Francorum and Other Contemporary Narratives," *Crusades* 3 (2004): 77–99.

127 Harari, "Eyewitnessing," 99.

128 Elizabeth Lapina, "'Nec signis nec testibus creditor': The Problem of Eyewitnesses in the Chronicles of the First Crusade," *Viator* 38 (2007): 117–39. Similar views in: Simon John, "Historical Truth and the Miraculous Past: The Use of Oral Evidence in Twelfth-Century Latin Historical Writing on the First Crusade," *English Historical Review* 130 (2015): 263–301.

The Artificiality of the "Battle-piece": The Role of the Author's Education

Rhetorical education can be considered a second reason for the artificiality of "battle-pieces." Rhetoric was considered to be a fundamental component of secondary education in Byzantium, and the key teaching tool of the teachers, at several levels of the *ἐγκύκλιος παιδεία* between the ages of eleven and sixteen,[129] were "preliminary exercises," known as *progymnasmata*, aimed at training the student to compose short texts on themes drawn from ancient Greek mythology, the origins of popular sayings, or eulogies in praise of historical or mythological characters, or comparing persons and events of opposing qualities.[130] As further (educational) progress depended on the student's ability and ambition, his/her parents' means and the availability of a teacher, any kind of career in the higher levels of state bureaucracy required proficiency in this kind of literary education. Yet, even writers such as the late eleventh-century Kekaumenos, a man of modest provincial background, would have been aware of some basic *progymnasmata*,[131] let alone well-educated judges like Attaleiates or Skylitzes.

It was Hermogenes of Tarsus (ca. 160–230 AD) who emerged as, perhaps, the most influential rhetorical theorist of Late Antiquity, during the time of the so-called Second Sophistic. Amongst the most authoritative rhetorical handbooks in the field were Aphthonios's *Progymnasmata* (later fourth century), Hermogenes's *Peri Staseon* ("On Issues") and *Peri Ideon* ("On Ideas"), and Menander of Laodicea's epideictic treatises.[132] Hermogenes's and Aphthonios's *Progymnasmata* introduced students to the basics of rhetorical composition by training them in a variety of discursive habits and methods: narration (*On Fable* and *On Narrative*); short memorable phrases (*On Chreia* and *On Maxim*); praise and blame (*On Encomium*: *epainos* and *psogos*); an imitation of the character of the speaker (*On Ethopoeia*); description (*On Ekphrasis*). The relevant part for our purposes is the *description/ekphrasis*, which is defined by Hermogenes as the "descriptive speech, as they say, vivid and bringing what is being shown before the

129 Athanasios Markopoulos, "Education," in *Oxford Handbook of Byzantine Studies* (Oxford: Oxford University Press, 2009), 785–95.

130 Hunger, *Βυζαντινή λογοτεχνία*, 1:160–96. See also: Craig A. Gibson, "Learning Greek History in the Ancient Classroom: The Evidence of the Treatises on *Progymnasmata*," *Classical Philology* 99 (2004): 103–29.

131 Charlotte Roueché, "The Rhetoric of Kekaumenos," in *Rhetoric in Byzantium*, ed. Elizabeth Jeffreys (Aldershot: Ashgate, 2003), 23–37.

132 Stratis Papaioannou, "Theory of Literature," in *The Oxford Handbook of Byzantine Literature*, ed. Stratis Papaioannou (New York: Oxford University Press, 2021), 76–109 (especially: 78–88); Elizabeth Jeffreys, "Rhetoric," in *Oxford Handbook of Byzantine Studies* (Oxford: Oxford University Press, 2009), 828–33; Jeffreys, "*Rhetoric in Byzantium*," in *A Companion to Greek Rhetoric*, ed. Ian Worthington (London: Blackwell, 2007), 166–84 (especially: 168–71); George A. Kennedy (ed. and trans.), *Progymnasmata, Greek Textbooks of Prose Composition and Rhetoric* (Atlanta: Society of Biblical Literature, 2003); Hugo Rabe and George A. Kennedy (ed. and trans.), *Invention and Method in Greek Rhetorical Theory: Two Rhetorical Treatises from the Hermogenic Corpus* (Atlanta: Society of Biblical Literature, 2005).

eyes."[133] He also adds that two critical virtues of the *ekphrasis* are "clarity and vividness; for the expression should almost create seeing through the hearing." And because war, battles, and sieges have always been an attractive topic for a rhetorical *ekphrasis*, Hermogenes adds that "if we were speaking an ekphrasis of a war, first we shall mention events before the war: recruiting the soldiers, the expenditures, the fears; then the attacks, the slaughter, the deaths; then the victory trophies; then the paeans of the victors and the others' tears and slavery."

The Greek sophist and rhetorician Libanius of Antioch (314–393 AD) expanded upon Hermogenes's instructions concerning the description of an infantry battle, emphasizing accounts of those killed ("one man having done nothing, lay dead on his shield, while another lay dead on the ground"), the wounds suffered ("and one man's hand was cut off, while another's eye was knocked out, and one lay struck in the groin, while someone tore open another's stomach"), and the weapons employed ("when the arrows had been used up by the archers, the battle became hand-to-hand, and there was much fighting with spears and clashing of arms").[134] The gory elements described by Libanius were common in Homeric epic poetry,[135] and they often feature in the Late Antique and Middle Byzantine historiography, illustrating that the emphasis on wounds and blood forms part of a wider interest in graphic violence in these periods.[136]

The sophists encouraged several literary models for writers to follow, the most important of whom were Homer and Thucydides.[137] Throughout the ages, Homer has offered soldiers numerous models of courage and leadership, whilst historians can look to the *Iliad* or to *The Peloponnesian War* on how to structure his "battle-pieces." There are numerous episodes depicting champions in heroic duels in the Middle Byzantine historiography that can trace influence back to the *Iliad*[138] when a warrior's virtue

---

133 Kennedy, *Progymnasmata*, 86.

134 Libanius's *Progymnasmata: Model Exercises in Greek Prose Composition and Rhetoric*, translated with an introduction and notes by Craig A. Gibson (Atlanta: Society of Biblical Literature, 2008), 429–31.

135 Simon Hornblower, "Warfare in Ancient Literature: The Paradox of War," in *The Cambridge History of Greek and Roman Warfare: Volume 1, Greece, The Hellenistic World and the Rise of Rome*, ed. Philip Sabin, Hans van Wees, and Michael Whitby (Cambridge: Cambridge University Press, 2007), 22–53 (especially 24, 48); Schein, *The Mortal Hero*, chapter 3.

136 Compare the description of the assassination of Nikephoros Phokas in Talbot and Sullivan, *The History of Leo the Deacon*, 138, and Homer, *Iliad*, 16:324. See also: Walter Pohl, "Perceptions of Barbarian Violence," in *Violence in Late Antiquity: Perceptions and Practices*, ed. Harold Allen Drake (New York: Routledge, 2016), 15–26; John W. Birkenmeier, *The Development of the Komnenian Army, 1081–1180* (Leiden: Brill, 2002), chapter 8; Whately, *Combat, Culture, and Didacticism*, 160–68.

137 Robert Browning, "Homer in Byzantium," *Viator* 6 (1975): 15–33; Diether Roderich Reinsch, "Byzantine Adaptations of Thucydides," in *Brill's Companion to Thucydides*, ed. Antonis Tsakmakis and Antonios Rengakos (Leiden: Brill, 2006), 755–78.

138 Poems about heroes killing and dying for their own imperishable glory are very old in the Indo-European literary tradition: Seth L. Schein, *The Mortal Hero*, An Introduction to Homer's *Iliad* (Berkeley: University of California Press, 1984), chapter 1. For the place of these episodes in the "Homeric" construction of battle-pieces: Hans van Wees, "Homeric Warfare," in *A New Companion to Homer*, ed. Ian Morris and Barry Powell (Leiden: Brill, 1996), 673–74.

(*arete*)[139] was constantly assessed according to the prowess and courage he displayed. Kaldellis has linked the reignited Byzantine interest in Homer to the Middle Byzantine military aristocracy's need for heroic models, adding that "Homer was in the air, fuelling a shift in values among rulers and writers."[140]

To give a handful of examples, like Leo the Deacon's description of the dash made by Anemas—the commander of the imperial bodyguard at the Battle of Dorystolon (971)—against the second-in-command of the Rus' army; or the dramatic duel between Bardas Skleros and Bardas Phokas, recorded in Skylitzes's *Synopsis*; or Niketas Choniates's description of the duel between a Byzantine and an Armenian during John II Komnenos's campaign in Cilicia in 1137–1138, when the Byzantine split the "Hectorian" shield of his opponent in a violent clash of arms.[141] Bryennios's account also highlights that had Romanos been allowed to engage with the Turks in a "Homeric/chivalric" hand-to-hand fight (Greek: *εις χεῖρας ἐλθεῖν*), he would have emerged triumphant, but the Turks used "cunning" (i.e., hit-and-run) tactics to avoid that and, thus, denied Romanos the opportunity for a fair "Homeric *ἀγών*."[142]

There are various reasons as to why Byzantine historians "borrowed" from classical authors: to embellish their texts, to show off their classical education, to give credence to their accounts, and to showcase the cultural and literary models they wished to emulate.[143] Here, I give a handful of examples of classicizing historians from the Early and Middle Byzantine periods who used ancient literary models in their "battle-pieces."Cameron "used (and will continue to use) the term 'Thucydidean history' in relation to Procopius's *Wars*, because it does usefully denote a whole set of external characteristics to which Procopius largely conforms."[144] Procopius's use of Thucydidean narrative strategies is shown to be particularly evident in his accounts of battle, and, Kaldellis applied de Romilly's model to the Battle of Mammes in the *Vandal Wars* to deduce that Procopius's technique of describing battle using speeches made by opposing leaders as "literary vehicles of military analysis" is "Thucydidean through and through."[145] Thucydides also serves as a literary model for describing naval battles, cavalry encoun-

139 Whately, *Combat, Culture, and Didacticism*, 177–81.

140 Anthony Kaldellis, "Classical Scholarship in Twelfth-Century Byzantium," in *Medieval Greek Commentaries on the Nicomachean Ethics*, ed. Charles Barber and David Jenkins (Leiden: Brill, 2009), 21–22.

141 Sinclair, "War Writing," 320–28; Talbot and Sullivan, *The History of Leo the Deacon*, 43; Maniati-Kokkini, "Η επίδειξη ανδρείας," 239–59; Whately, *Combat, Culture, and Didacticism*, 169–71; Martin van Creveld, *Wargames: From Gladiators to Gigabytes* (Cambridge: Cambridge University Press, 2013), 39–47.

142 Bryennios, *Ύλη Ιστορίας*, 90–91; Neville, *Heroes and Romans*, 97–99.

143 Neville, "Why Did the Byzantines Write History?," 272–76; Sinclair, "War Writing," 13–21.

144 Averil Cameron, *Procopius and the Sixth Century* (London: Routledge, 1996), 32–45.

145 Kaldellis, *Procopius of Caesarea*, 29–31; Jacqueline de Romilly, *Histoire et raison chez Thucydide* (Paris: Belles-Lettres, 1956), especially 138–50. See also: Philip Rance, "Wars," in *A Companion to Procopius of Caesarea*, ed. Mischa Meier and Federico Montinaro (Leiden: Brill, 2022), 70–120; in the same volume: Anthony Kaldellis, "The Classicism of Procopius," 339–54.

ters, and sieges, and his narrative about the naval battle at the Great Harbour of Syracuse and the siege of Plataea are well-known to have served as a "literary inspiration" for many centuries to come.[146]

Already in the late nineteenth century, Braun affirmed the Procopian mimeses of Thucydides in his narrative of the plague of Constantinople in 542 as an allusion to the plague of Athens, and also in his accounts of the Roman sieges of Amida, Edessa, and Rome that "borrow" from Thucydides's narrative of the siege of Plataea.[147] Since Procopius was also intimately acquainted with the realities of siege warfare, he inevitably developed a great interest in Thucydides's siege narratives, which can explain the numerous textual parallels between Procopius's description of the siege of Naples and Thucydides's siege of Plataea that have been put forward by Pazdernik.[148] Whately's exemplary study also shows the importance of the role of the Homeric *arete* in the explanations of many *Gothic Wars*' combats, while he also points out that Procopius's famous comparison between contemporary horse-archers and their Homeric forebears in the preface to the *Gothic Wars* served as a medium for Procopius to advocate the tactical superiority of the horse-archers, and to argue that the Roman soldiers have this *arete* for which they have been censured.[149]

Theodosios the Deacon's *Capture of Crete*, a panegyric written to commemorate the capture of the island by Nikephoros Phokas in 961, frequently alludes to Homer and is evidently inspired by the *Iliad* in his description of heroic duels and epic battles.[150] Furthermore, writing about the lost common source used by Leo the Deacon and Skylitzes for their account of John Tzimiskes's Balkan campaign of 971 AD against the Rus, Kaldellis argues that this source featuring heroic battles, sieges, and speeches by both the Byzantine and the Rus' leaders, draws upon and compares Tzimiskes with glorious figures from the Roman past (Plutarch's *Camillus*). Also according to Kaldellis, the unknown author's literary elaboration of the belief that the battle (of Dorystolo) was won (partly) through saintly intervention evokes strong parallels Dionysios of Halikarnassos's account of the intervention of the Dioskouroi at the Battle of Lake Regillus (499 or 496 BC), a legendary Roman victory over the Latin League. The elaboration offers

---

146 Conor Whately, "Thucydides, Procopius, and the Historians of the Later Roman Empire," in *The Oxford Handbook of Thucydides*, ed. Ryan K. Balot, Sara Forsdyke, and Edith Foster (New York: Oxford University Press, 2017), 691–707.

147 Hermann Braun, *Procopius Caesariensis quatenus imitatus sit Thucydidem* (Erlangen: Junge, 1885), 31–34 and 58 (for the plague), and 47–61 (for the sieges).

148 Charles F. Pazdernik, "Procopius and Thucydides on the Labors of War: Belisarius and Brasidas in the Field," *Transactions of the American Philological Association* 130 (2000): 149–87; Pazdernik, "Belisarius' Second Occupation of Rome and Pericles' Last Speech," in *Shifting Genres in Late Antiquity*, ed. Geoffrey Greatrex and Hugh Elton (Farnham: Ashgate, 2015), 207–18. See also: Whately, "Thucydides, Procopius, and the Historians," 698–700.

149 Whately, *Combat, Culture, and Didacticism*, 177–88.

150 Théodose Le Diacre, *La prise de la Crète (960–961)*, trans. by René-Claude Bondoux and Jean-Pierre Grélois (Paris: Association des Amis du Centre d'Histoire et Civilization de Byzance, 2017), 21–28; Marc D. Lauxtermann, *Byzantine Poetry from Pisides to Geometres, Texts and Contexts* (Vienna: Austrian Academy of Sciences, 2019), 2:41–49.

"new models for thinking about the Byzantine triumph, and about the intercession of Christian military saints on the field of battle."[151]

It has been acknowledged for many decades that Leo the Deacon adopted literary models for his battle and siege descriptions from Homer, Thucydides, and from the sixth-century historian Agathias.[152] Treadgold has shown that Leo's allusions to Procopius and Agathias resemble Procopius's own allusions to Herodotus and Thucydides. Examples are found in Leo's description of the Athenians disastrous Byzantine expedition to Sicily that recalls Thucydides's own accounof the doomed enterprise; the harlot who mocked Phokas's troops from the walls of Chandax as a reference to Procopius's story of the courtesans who mocked the Persians from the walls of Amida;[153] or Leo's reference to careless raids while besieging a city that alludes to Maurice's *Strategikon* and Thucydides's description of the siege of Plataea.[154] Leo's "weather metaphors" when writing about bombardments of arrows and stones also reflect the literary technique of Agathias and Homer.[155] Finally, Michael Sjuzjumov has shown that Leo the Deacon's description of the undermining of the wall of Chandax by Phokas's troops in 961 was copied (almost verbatim) from Agathias's description of Narses's siege of Cumae.[156]

Bryennios shows great familiarity with Homer and ancient Greek literature and skilfully adds several proverbs to his work, with his military vocabulary being reminiscent of Xenophon.[157] Scholars have also acknowledged the "Homeric air" in Anna's *Alexiad*, and Buckler commented several decades ago that "[Anna's] descriptions of real battles were coloured by deep-seated memories of the siege of Troy," adding that in Anna's heroic models "her husband stands out among his contemporaries as did Achilles whom 'Homer hymned'; Turkish tactics are once actually contrasted with those of the *Iliad*."[158] Ruth Macrides has further suggested that Anna's focus on military affairs placed her firmly in the tradition of classicizing historians, and, moreover, in the epic, Homeric style that characterized twelfth-century descriptions of warfare. Ljubarskij has discussed the consistency with which the *Alexiad* is constructed around its central figure (Alexios) in the manner of Homeric epic.[159]

---

151 Kaldellis, "The Original Source," 35–52.

152 Treadgold, *Middle Byzantine Historians*, 243–44; Karpozilos, *Βυζαντινοί Ιστορικοί και Χρονογράφοι*, 2: 492–501; Hunger, *Βυζαντινή λογοτεχνία*, 2:184–85.

153 Treadgold, *Early Byzantine Historians*, 189, 217–18.

154 Denis Sullivan, "Tenth Century Byzantine Offensive Siege Warfare: Instructional Prescriptions and Historical Practice," in *Byzantium at War*, 179–200 (especially 189).

155 Sinclair, "War Writing," 58 (also nos. 302–5).

156 Reference in: Sullivan, "Tenth Century Byzantine Offensive Siege Warfare," 181n9. See also: Denis F. Sullivan, *Siegecraft: Two Tenth-Century Instructional Manuals by "Heron of Byzantium"* (Washington, DC: Dumbarton Oaks , 2000), 18–21.

157 Treadgold, *Middle Byzantine Historians*, 352n52; Neville, *Heroes and Romans*, 39–43.

158 Buckler, *Anna Comnena*, 198–99.

159 Ruth Macrides, "The Pen and the Sword: Who Wrote the *Alexiad*?," in *Anna Komnene and Her Times*, ed. Thalia Gouma-Peterson (New York: Garland, 2000), 63–81. In the same volume: Jakov Ljubarskij, "Why is the *Alexiad* a Masterpiece of Byzantine Literature?," 169–85. See also: Penelope

Because literary *kainotomia* (*καινοτομία*: novelty) was generally shunned in Byzantium, every literary work, including military manuals, aimed at an educated reader echoed literary models from the past. Yet, what historians have coined as the Byzantine *mimeses* (*μίμησις*: imitation) is no longer considered to be a "slavish imitation" of ancient models, but rather an innovative paraphrase of their sources to fit their period and context, so that "the story remains the same, but now becomes another."[160]

The Limitations of the Narrative Sources for Military History

So, to return to the main topic of this section, what exactly are the limitations of medieval sources and what danger do they pose for a modern researcher? In many cases, several of the clerical sources give an incomplete narrative of battles, sieges, or entire campaigns, simply because reporting these events in a manner similar to a modern war correspondent was not the objective.[161] It has to be stressed, however, that it is undoubtedly an oversimplification to argue that all clerics were ignorant of military affairs, keeping in mind that Orderic Vitalis and William of Poitiers are two of the foremost sources for Anglo-Norman military history,[162] while Amatus of Montecassino and Geoffrey Malaterra were both monks who wrote detailed narrative accounts of the conquest of southern Italy and Sicily by the Normans in the eleventh century.[163] Their Muslim counterparts were also decidedly non-professional military historians, and we should not expect specialized military insight from their works.[164] But even secular historians with

Buckley, *The* Alexiad *of Anna Komnene: Artistic Strategy in the Making of a Myth* (Cambridge: Cambridge University Press, 2014), chapter 2.

160 Ingela Nilsson, "The Same Story, but Another, A Reappraisal of Literary Imitation in Byzantium," in *Imitatio—Aemulatio—Variatio, Akten des internationalen wissenschaftlichen Symposiums zur byzantinischen Sprache und Literatur (Wien, 22.-25. Oktober 2008)*, ed. A. Rhoby and E. Schiffer (Vienna: Österreichischen Akademie der Wissenschaften, 2010), 195–208; Reinsch, "Byzantine Adaptations of Thucydides," 755–78; Kaldellis, *Procopius*, 24–45; Michael Bartusis, "The Function of Archaizing in Byzantium," *Byzantinoslavica* 56 (1995), 271–78; Guyla Moravcsik, "Klassizismus in der byzantinischen Geschichtsschreibung," in *Polychronion: Festschrift Franz Dölger zum 75 Geburtstag*, ed. P. Wirth (Heidelberg: Winter, 1966), 366–77; Cameron, *Procopius*, 32–45. The classical study on the topic is by Herbert Hunger, "On the Imitation (ΜΙΜΗΣΙΣ) of Antiquity in Byzantine Literature," *Dumbarton Oaks Papers* 23/24 (1969/1970), 15–38.

161 Verbruggen, *The Art of Warfare*, 10–11. On the value of saints' lives for the study of military history: John France, "War and Sanctity: Saints' Lives as Sources for Early Medieval Warfare," *Journal of Medieval Military History* 3 (2005): 14–22.

162 William of Poitiers was a cleric, but he had been a knight in his earlier life and had acquired an excellent classical education at Poitiers. Yet, his descriptions of battles are full of artificial elements because of his classicizing desire to show off his knowledge of Latin. Emily A. Winkler, "The Norman Conquest of the Classical Past: William of Poitiers, Language and History," *Journal of Medieval History* 42 (2016): 456–78.

163 I examine the value of both these sources for the history of warfare in the Mediterranean in: Georgios Theotokis, "Greek and Latin Sources for the Norman Expansion in the South: Their Value as 'Military Histories' of the Warfare in the Mediterranean Sea," in Theotokis, ed., *Warfare in the Norman Mediterranean*, 11–33.

164 Hillenbrand, *The Crusades*, 432.

no military background found it "tiresome" going into detail as far as battles, sieges, etc., are concerned. This is clear as we read in the *prooimion* of the work of a twelfth-century Byzantine judge and bureaucrat who became a monk later in his life:

> By the majority, not to say perhaps by all of those who have read the compositions of those histories, these things are regarded as tiresome and off-putting because they both require a surfeit of leisure and because, even if some who apply themselves to the histories manage to be so fortunate, the effort expended on them turns out vain, since the lengthy expositions about alignments and wars and the soldiers' orders-of-battle and of the rest of such things elude their memory, and, in truth, the public speeches and the lectures verge on the useless for those who apply themselves to what is being recounted.[165]

In relation to the last point, we should bear in mind that certain of our authors would not have been experienced in military affairs, a fact that always bears the risk of inaccurate or erroneous reports of events. A good example of this is the report of Emperor John Tzimiskes's feigned retreat during the final engagement with the Rus' outside Dorystolon (July 971). Skylitzes describes this manoeuvre in every detail, while Leo the Deacon merely mentions that the Byzantine troops retreated at head-long speed.[166] Nonetheless, we should not rush to dismiss Leo the Deacon's work as useless for military history; his vocabulary for military units, equipment, ships, and so on, may be archaic, but his descriptions often reflect what the military manuals recommend, and his narrative is extremely valuable for reconstructing the military events of his period.[167] Likewise, Albert of Aachen was a canon of the church of Aachen and he gathered oral and written testaments of participants in the Crusade for his *Historia expeditionis Hierosolymitanae*. His work remains an important source on the First Crusade and the history of the kingdom of Jerusalem until 1120 and is the sole document on the People's Crusade of 1096.[168] Yet, his account has some brilliant insights into the First Crusade, the march, looting, fighting, being taken prisoner, etc., and, although, little is known about his life, he probably never visited the Holy Land.

The chroniclers' works are also affected by exaggeration depending on their biases and sympathies, and I believe I have made this abundantly clear thus far with regards to the pro- and anti- Romanos sources at our disposal for the events leading up to, during, and following the Battle of Manzikert. Other times, their narrative is embellished with religious elements owing to their religious convictions, as very often they tend to ascribe victory or defeat to divine judgement and intervention, or, in the (rare) case of

165 Banchich, *The History of Zonaras*, 24.

166 Skylitzes, *Synopsis*, 291; Leo the Deacon, *History*, 196.

167 Talbot and Sullivan, *The History of Leo the Deacon*, 36–47; Treadgold, *Middle Byzantine Historians*, 244–45.

168 Albert of Aachen, *Historia Ierosolimitana: History of the Journey to Jerusalem*, ed. and trans. Susan B. Edgington (Oxford: Clarendon, 2007).

Caesar, to *fortuna*.[169] The motif of divine aid for individual heroes has been common since Agamemnon's snub of Achilles: "If you are very strong, it was a god, I think, who gave you this gift. Go home with your ships and your companions and lord it over the Myrmidons; for you I care not."[170] Describing the military achievements of his "hero" Botaneiates against the Patzinaks in Thrace in the early 1050s, Attaleiates notes that "to walk on foot for so many days while fighting and not have any rest at night, well, no one would have ever heard of such a feat if it had not been accomplished in the case of this man through supernatural assistance."[171] For Skylitzes, John Tzimiskes also displayed "supernatural courage in so small a body,"[172] yet more appropriate for my analysis here is John's narrative for the battle at Dorystolon (971), where "the Romans are said to have benefited from the enhanced supernatural aid at that time, for a storm arose in the south...And a man appeared to the entire Roman army mounted on a white horse they say he was one of the [two] gloriously victorious martyrs named Theodore, for the emperor always used [the icons of] these martyrs as allies and protectors against the foe."[173] For God's "chosen people" of Romania, military success (not just over infidels but over every enemy of the empire)[174] was associated with divine blessing and, conversely, military defeat with the withdrawal of blessing because of sin.[175] However, it is fair to say that not everyone was convinced about God's intervention to decide a battle, as we read in Procopius's *Vandal Wars*: "From then on, I am unable to say what happened to Gelimer that, having the victory in his hands, he willingly gave it up to the enemy. Unless one ought to refer foolish actions also to God."[176]

Muslim authors of medieval historical sources were more often than not religious scholars who had devoted their lives to the in-depth study of the Quran before becoming interested in the writing of history, which they saw through the prism of faith. For them, God's will is the ultimate cause of all events, the tools He preferred were the actions of elite individuals, and the standard to which He held his creatures had been made clear in His law. Therefore, with God in the epicentre of history, Muslim historians had to exemplify known truths and to teach lessons by describing them.[177] We read Usama's declaration: "Victory in warfare is from Allah (blessed and exalted is he!) and is not due to

169 "Fortune accomplishes much, not only in other matters, but also in the Art of War." C. Julius Caesar, *Gallic War*, 6:30.

170 Homer, *Iliad*, bk. 1, 176–80; Marincola, *Authority and Tradition*, 2078.

171 Attaleiates, *History*, 75.

172 Skylitzes, *Synopsis*, 296.

173 Skylitzes, *Synopsis*, 292.

174 Yannis Stouraitis, "Using the Bible to Justify Imperial Warfare in High-Medieval Byzantium," in *The Bible in Byzantium: Appropriation, Adaptation, Interpretation*, ed. Claudia Rapp, Andreas Külzer, *Journal of Ancient Judaism Supplements* 25 / Reading Scripture in Judaism and Christianity 6 (Göttingen: Vandenhoeck and Ruprecht, 2019), 89–106.

175 Riedel, *Leo VI*, 162–81.

176 Procopius, *Vandal Wars*, bk. 3, chap. 19, 25–29.

177 Robinson, *Islamic Historiography*, 129–34.

organisation and planning, nor to the number of troops and supporters."[178] When Ibn al-Athīr wrote about the defeat inflicted on Baldwin I in 1102 by the Egyptians, he added "God gave victory to the Muslims,"[179] while his account of the Muslim operations against the Sicilian fleet besieging Alexandria in 1174 is peppered with phrases such as: "God sent down His aid to them and His signs became manifest" and "Thus God delivered the Muslims from their wickedness."[180] Ibn al-Athīr and Usama were very much Muslims of their time for whom the victories of the Muslims over the infidels are a manifestation of God's will: to make truth triumph over error, to reconquer lost territories, and to ensure the unity of *Dār al-Islām*.[181] Indeed, what a striking contrast to Procopius's *dictum*, placed into the mouth of Leo Phocas by Leo the Deacon: "For wars are usually won not so much by a pitched battle as by cautious planning,"[182] or Leo VI's claim in the *prooimion* to his *Taktika*: "it is not true, as some inexperienced persons may hold, that wars are decided by a multitude of men and courage, but by the favour of God and by generalship and discipline."[183]

Therefore, in order to properly evaluate our sources as military historians we must know the socioeconomic and religious background, life and career, and the specific cultural context in which the authors wrote their narratives. We must also become cognizant of how they "understood" battle, and the literary models that underscored their writing, certainly not whether their descriptions fit our modern understanding. Hence, in the final section of this chapter I will ask how accurate and detailed the descriptions of pitched battles and clashes between enemy units during the Byzantine campaigns of 1068 to 1071 in the East are. How far do the narratives permit the accurate reconstruction of chains of events, especially in regard to the battlefield manoeuvres of armies in action?

A Comparison of the "Battle-Pieces" for the Byzantine Campaigns of 1068 to 1071

Attaleiates and Skylitzes explain how the "miserable and downcast, and angry" imperial troops posted in and around Melitene were put to flight in the early summer of 1067 by Turkmen troops who were "shooting from a great distance" in their typical nomadic fashion. It was a clash that opened the way for the invaders to proceed west-

178 Usamah ibn Munqidh, *Memoirs of an Arab-Syrian Gentleman, or, an Arab Knight in the Crusades: Memoirs of Usāmah ibn-Munqidh (Kitāb al-i'tibār)*, trans. P. Khuri Hitti (New York: Columbia University Press, 1929), 177.

179 D. S. Richards (trans.), *The Chronicle of Ibn al-Athīr for the Crusading Period from al-Kāmil fī'l-ta'rīkh (Part 1)* (New York: Routledge, 2016), 61.

180 D. S. Richards (trans.), *The Chronicle of Ibn al-Athīr for the Crusading Period from al-Kāmil fī'l-ta'rīkh (Part 2)* (New York: Routledge, 2016), 229–30.

181 Françoise Micheau, "Ibn al-Athīr," in *Medieval Muslim Historians and the Franks in the Levant*, 52–83.

182 *History of Leo the Deacon*, 73.

183 *The Taktika of Leo VI*, prologue, 8.

ward into the heartland of Kappadokia and to attack Kaisarea.[184] Regarding Romanos's campaign in the main operational theatre of Syria in 1068, both sources note that, following the pillage of the rural environs of the city of Aleppo in the first week of October, the emperor marched for three days to Hierapolis (modern Manbij).[185] I will discuss in the following chapter the rare topographical descriptions they include of the plains around Hierapolis, where they emphasize the flatness of the ground that made it ideal for cavalry action, while also noting the irrigation canals to the west of the city that posed an obvious obstacle for mounted units.[186] Yet, the description of the battle outside Hierapolis only highlights the basic battle tactics of the Arab and Turkmen defenders, in that they "gave the impression of mounting an attack, [and] they did not dare close with the Romans in hand to hand fighting, contenting themselves with skirmishing from afar with projectiles and forays." [187] Our sources keep silent on how, exactly, Romanos took on these attacks, and we only learn the important outcome of the battle in that the Byzantines seized the town and several towers of the ramparts and, following an intense bombardment of the citadel, they brought about its surrender under terms.

Attaleiates and Skylitzes are somewhat more detailed when describing the joint attack by the Mirdasid *emir* of Aleppo, Mahmūd bin Naṣr, and a Turkoman contingent that followed Romanos's conquest of the lower town of Hierapolis in November (1068) which resulted in a defeat of the Byzantine units posted to defend the area between the imperial camp and the town. They write of the "Saracen" cavalry's attempts to harass and dislodge the imperial units from their positions (in the camp?) so that they would "ride out after them"; a typical tactic used by these Arab mounted units was to throw missiles from afar (Greek: *ἀκροβολισμοὺς δὲ ποιησάμενοι*)[188], which fits with the well-known tactic of the *Arabitai* described in Nikephoros Phokas's *Praecepta Militaria* (ca. 969).[189] In fact, the Byzantine hollow double (*διττή*) infantry square, which was devised in the mid-tenth century as a defensive formation and place of refuge for both infantry and cavalry, was created exactly to defend from the encircling and hit-and-ran tactics of these lightly armed and fast-moving Bedouins from northern Syria who attacked with javelins rather than with arrows.[190]

---

184 Attaleiates, *History*, 171; Skylitzes, *Continuation*, 73.

185 Attaleiates, *History*, 197–99; Skylitzes, *Continuation*, 87–89; Zonaras, *Επιτομή Ιστοριών*, 3:204–5.

186 Attaleiates, *History*, 203–5; Skylitzes, *Continuation*, 91.

187 Attaleiates, *History*, 199–203; Skylitzes, *Continuation*, 89–91.

188 The editors of Attaleiates's History correctly translate the verb ακροβολίζομαι as "to shoot from a distance," whether using arrows or javelins: Attaleiates, *History*, 204–5. The editor of the *Continuation of Skylitzes* writes, in my view incorrectly, that "they [Arabs] had rained arrows from afar": Skylitzes, *Continuation*, 90–91.

189 *Praecepta Militaria*, II.101–11, 28; "The Anonymous Byzantine Treatise on Tactics," in *Three Byzantine Military Treatises*, ed. and trans. George T. Dennis, §10 (Washington, DC: Dumbarton Oaks, 2008).

190 Theotokis, *Byzantine Military Tactics*, 219–21. For the employment of Bedouins in the armies of the Hamdanids and the Fatimids in the tenth century: McGeer, *Sowing the Dragon's Teeth*, 225–46.

Outside Hierapolis, these lightly armed mounted units eventually engaged in hand-to-hand fighting with the heavily armed cavalry tagma of the *stratelatai* and managed to route them, with the tagma vanishing from the record shortly after this clash.[191] Attaleiates reserves some vitriolic comments about the tactical attitude of the soldiers of the tagma of the *Scholai* who allegedly "even while witnessing the defeat of the others, gave no thought to coming to their assistance or doing anything else expected of soldiers, but rather they stayed fixed there." I mentioned earlier that this is a comment that greatly underestimates the tactical situation on the spot and gives no consideration to the advice of the military manuals; again, it is no coincidence that it comes from the pen of a judge rather than a military man. In a situation when the imperial army was forced onto the defensive, it deployed its units into a hollow double (*διττή*) infantry square with the heavy infantry forming a protective square "shield" around the cavalry. What was imperative was for all units to keep their formation intact in the face of repeated attacks, something that required immense levels of group- and self- discipline. Attaleiates notes that the Arabs eventually managed to lure some Byzantine units, "who rode out [from the camp?] against them, but did not then wheel around and retreat as in their earlier manoeuvres"; we should consider that the soldiers of the *Scholai* were simply following orders rather than their instinct, and stayed put and in formation rather than come out to assist their comrades.[192]

Also noteworthy is the hostile attitude of both our sources to the Armenian infantry in imperial service, who they blame for being undisciplined and for planning to defect to the enemy.[193] Ultimately, on November 20, 1068, the emperor organized a sortie that routed the enemy forces, although "the pursuit did not last a long time, the Romans missed a glorious opportunity and were deprived of a magnificent victory."[194] Was the Byzantine failure to pursue and inflict more casualties on the Arabs a result of miscommunication, failure of command, or fear/caution of being drawn into another ambush? We will never know.

For Romanos's campaign the following year (1069) that focused on re-gaining the strategic initiative in the Upper Euphrates valley, the emperor set out from Kaisarea and marched to Larissa (modern Mancilik) in the theme of Sebasteia. Attaleiates and Skylitzes describe a clash with the numerous forces of Turkmen that were pillaging the countryside around Larissa, in which the they had the advantage of holding the higher ground, and launched an attack on the imperial soldiers building a camp in the plains below. Although the rear guard that was protecting the *touldon* (the supply train) had not yet arrived, the main body of the imperial army immediately lined up for battle and

---

191 Attaleiates, *History*, 205–7; Skylitzes, *Continuation*, 91. On the stratelatai: Hans Joachim Kühn, *Die byzantinische Armee im 10. und 11. Jahrhundert: Studien zur Organisation der Tagmata* (Vienna: Fassbaender, 1991), 247–49. See also: Simeon Antonov, "Saint Theodore's Earthly Host? Tagma ton Stratelaton and Its Commanding Officers (late tenth–eleventh centuries)," *Byzantion* 92 (2022): 47–72.

192 I explain this in detail in: Theotokis, *Byzantine Military Tactics*, chapter 7.

193 Attaleiates, *History*, 207; Skylitzes, *Continuation*, 93.

194 Attaleiates, *History*, 209; Skylitzes, *Continuation*, 93.

marched against the enemy. The Turkmen were quickly repulsed from the vicinity of the camp, and the emperor put the remainder of the enemy to flight "with the Scythian [Uzes] mercenaries in the lead," eventually pushing them east as far as the banks of the Euphrates near Melitene.[195] It is noteworthy that (some of) the imperial units defending the camp were the tagmata of the *Lykaonians* and the *Hesperioi Arithmoi*, while a mercenary unit of "Franks" is also attested.[196]

In the following year (1070), when Kaisarea became the base of the (defensive) operations led by the *protoproedros* and *kouropalates* Manuel Komnenos, the imperial forces camped in the vicinity of Sebasteia. Although reduced in numbers due reinforcements being sent to the southeast border (Hierapolis), they engaged in a battle with the Turkmen led by the brother-in-law of the sultan of the Seljuks, Alp Arslan, the *emir* Chrysoskoulos (Arîsîaghî or Arîsîghî). Attaleiates and Skylitzes write that the Turks made use of their "usual tricks and stratagems" of feigned retreat to draw Manuel's forces out of their camp, but they report no other details about battle manoeuvres that day.[197] Eventually, because of the inexperience of their young commander, the Byzantines relaxed and unsuspectingly approached the abandoned enemy camp, and then they were attacked and suffered heavy losses, while Manuel himself was taken prisoner.

Paradoxically, shortly after his capture by the *emir*, Manuel succeeded in persuading Chrysoskoulos of the benefits of a personal alliance with the emperor, and the Turk (who obviously did not need much convincing due to his conflict with Alp Arslan at the time) defected to Constantinople with Manuel to hold negotiations with the emperor, who was thrilled, showering the Turk with honourary titles and gifts.[198] Attaleiates ascribes to him the eyes and physical appearance of a "Skythian," clearly to emphasize his "Otherness," but whether Attaleiates actually saw him in person in the capital (because he did not join Manuel's campaign in 1070) remains an open question. By describing him as a "Sskythian" and a "pygmy in height," whose "face was...ugly because this people are of Skythian ancestry and have inherited their depravity and deformity," Attaleiates was clearly using the literary *topos* of a grotesque image of the Turk in order to "demonize" the "Other".[199]

195 Attaleiates, *History*, 229–31; Skylitzes, *Continuation*, 99.

196 For the tagma of "Lykaonians," an archaic term probably used to designate the tourmai of the old theme of Anatolikon: Georgios Leveniotes, "Το θέμα/δουκάτο των Ανατολικών κατά το Β΄ήμισυ του 11ου αι." [The Theme/Ducate of Anatolikon in the Second Half of the Eleventh Century], Βυζαντιακά 25 (2005–2006): 33–101 (mainly 46–55). For the Hesperioi Arithmoi: Hélène Glykatzi-Ahrweiler, "Recherches sur l'administration de l'empire byzantin aux IX–XIème siècles," *Bulletin de correspondance hellénique* 84 (1960): 28.

197 Attaleiates, *History*, 255; Skylitzes, *Continuation*, 107. Bryennios does not refer to the feigned retreat tactics of the Turkmen but, rather, mentions the surprising attack upon the "unsuspected" and 'scattered" imperial troops: Bryennios, *Ύλη Ιστορίας*, 73–75.

198 Alexander D. Beihammer, "Defection Across the Border of Islam and Christianity: Apostasy and Cross-Cultural Interaction in Byzantine-Seljuk Relations," *Speculum* 86 (2011): 597–651 (for Chrysoskoulos, 608).

199 Attaleiates, *History*, 259–61. This episode is summarized in Skylitzes, *Continuation*, 109–11; sc 211. On these literary topoi about the peoples neighbouring the empire, see also Tacitus, *Germany and its Tribes*, §4–5; Jordanes, *The Gothic History*, §3–4. Ammianus's famous account of the Huns

Yet, Byzantium's superiority was confirmed once again by transforming Chrysoskoulos from a "barbarian Skythian" to a Roman dignitary when Emperor Romanos bestowed on him the honourary rank of *proedros.*

Although Attaleiates depicts Chrysoskoulos as an object of manipulation and an outsider with a grotesque appearance who had to be "cajoled into" Byzantine service (or servitude?), the narrative function of the Turk was completely different for Nikephoros Bryennios. In the *Hyle*, we read about a Chrysoskoulos who, when his friend Manuel Komnenos died, nearly died with him: one can read it as excessive emotion shown by a barbarian, but this is still emotion. For Bryennios, the emir displayed his ability not just to be a pawn in the games of others, but also his competence to take matters into his own hands, as can be glanced in his role in saving the rebellion of Nikephoros Botaneiates near Nicaea.[200] For Attaleiates, whose *History* has traditionally been dated to the year 1079/80, the Seljuks were mere nameless nomads from "the ends of the world" to be manipulated into the schemes of the Byzantine emperors, whilst Bryennios, writing in the second quarter of the twelfth century, portrays the Seljuks as men with reason who had, by then, become neighbours of Byzantium and had entered into the complex relations with the empire.[201]

For the fatal campaign of the summer of 1071, we read about the emperor's plan to concentrate his efforts on the re-capture of Manzikert and Khliat on the north shore of Lake Van, as these fortresses were in a very critical strategic position—something that is confirmed in Constantine VII's *De Administrando Imperio*—,[202] which could not be allowed to be used by the Turks as bases for further penetration in the west. With the Greek sources being categorized as pro- and anti-Romanos as we have seen already, Attaleiates and Skylitzes are inclined to excuse the emperor's decision to dispatch almost half of his troops, including much of the effective cavalry, to support those already operating under Roussel around Khliat. Both historians go as far as to write that "dividing his [Romanos] forces was not irrational or lacking in strategic calculation. It was fate, or rather the wrath or design of God beyond our understanding, that turned the result to the contrary."[203]

Unsurprisingly, Bryennios provides a different view of the background behind the fateful decision to launch a major strike towards Lake Van, reporting that both the respected Georgian *magistros* Joseph Trachaneiotes and Nikephoros Bryennios "the Elder" sought to convince Romanos not to take to the field against the Turks but rather to wait for them in Theodosiopolis so that Alp Arslan would "be forced to clash with the Romans where it is in their [Romans] interest to make war. But they [Trachaneiotes

is heavily infused with ethnographic stereotypes: *Ammianus Marcellinus*, 31. 2. 1–12. As is his less well-known account of the "Saracens," 14. 4. 3.

200 Bryennios, *Ὕλη Ἱστορίας*, 235–37.

201 For more on the reflection of the Turkish identity in the Byzantine sources: Roman Shliakhtin, "From Huns into Persians: The Projected Identity of the Turks in the Byzantine Rhetoric of Eleventh and Twelfth Centuries," (PhD diss., Central European University, 2016), particularly chapters 2 and 3.

202 *De Administrando Imperio*, 204 (§44).

203 Attaleiates, *History*, 275; Skylitzes, *Continuation*, 117; Zonaras, *Ἐπιτομή*, 213.

and Bryennios] were under the impression that they were talking to a deaf man...more attention was paid [by Romanos] to the flatterers and not to those who advised the right thing."[204] Bryennios's "I told you so" moment with the benefit of hindsight, coupled with his criticism of the emperor's strategic decision, could not be more palpable here.

On Wednesday, August 24 (1071), when Bryennios "the Elder" was dispatched by the emperor with a relatively small force to support a foraging party from Manzikert that was being attacked by the Turks, he, in turn, found himself lured into ambush and the danger of being encircled by the enemy, so he had to withdraw while requesting reinforcements. Interestingly, Attaleiates and Skylitzes noticed that "in this confusing mêlée...these Turks were more courageous than the others we had experienced as they charged more boldly and stood up to their opponents in hand-to-hand combat."[205] It began to dawn on the Byzantine forces that they were not facing Turkmen but a more disciplined and well-trained Seljuk force. And when the emperor sent the *doux* of Theodosiopolis, Nikephoros Basilakes, with a larger detachment of "local soldiers" (i.e., Armenians) to repel the Turkish forces and secure supplies for the main army, he acted impetuously, allowing his unit to pursue the enemy too far, themselves being lured into an ambush. Cut off from any reinforcements, Basilakes and many of his men were captured.[206] Critically, Bryennios reports that Basilakes was the first to be dispatched to intercept the Turkish party that was harassing the foragers, and Bryennios "the Elder" was sent only later "to save" his unfortunate and impetuous namesake. Apparently, this (rather unlikely) version of events was designed to keep Bryennios's reputation untarnished for posterity.[207]

Only after the aforementioned clashes did Romanos realize that the sultan's troops had arrived in the area of Manzikert. By this time the Seljuk army had properly established camp in the foothills of the Süphan Daği, thus dominating the plateau that stretched south and southeast of Manzikert. Much to the emperor's consternation, the Turks (initially) refused to give battle, prompting Attaleiates to launch another tirade of stereotypes against them, "for the Turks are wicked by nature and masters of deceit; they accomplish everything by trickery and unabashed reversals."[208] Interestingly, Skylitzes's account omits these stereotypes and simply notes that "He [Romanos] took up a position on some hills until evening, but since he could see no opposing force, he returned to the encampment."[209]

That same evening (August 24), the agile Turkish horsemen eventually launched a raid against the Byzantine camp, and "with inarticulate howling and shooting of arrows they rode around terrifying them,"[210] succeeding in driving the Uzes mercenaries in

---

204 Bryennios, *Ὕλη Ιστορίας*, 78–83.

205 Attaleiates, *History*, 281; Skylitzes, *Continuation*, 117.

206 Attaleiates, *History*, 283; Skylitzes, *Continuation*, 117–19; Zonaras, *Επιτομή*, 213.

207 Bryennios, *Ὕλη Ιστορίας*, 85.

208 Attaleiates, *History*, 285.

209 Skylitzes, *Continuation*, 119.

210 Attaleiates, *History*, 285; Skylitzes, *Continuation*, 119; Zonaras, *Επιτομή*, 215.

imperial service to seek refuge inside the camp, where the imperial army faced further harassment from the enemy during the night (August 24/25). The following morning (Thursday, August 25), the Turks attempted to capture the opposite banks of the river that flowed nearby the imperial camp, in order to cut off the Byzantines' water supply, but we have no details of the small-scale battle between a Byzantine infantry detachment and the Turks.[211]

On the eve of the main battle on Friday, August 26, Bryennios was the only one who wrote that the left wing of the imperial army consisted mainly of the western tagmata under Nikephoros Bryennios "the Elder," and that the right wing included the units of Kappadokia and the rest of Asia Minor under Theodore Alyates. The emperor was apparently in command of the centre, placing Andronikos Doukas in charge of the rear guard and in command—according to Bryennios—of "the tagma of the *Hetairiae* and that of the nobles" (*τάς τε των έταίρων τάξεις έχων και τας των άρχόντων*).[212]

I will not go into great detail here about the phases of the battle that followed. Our sources make it clear that the Byzantines quickly sought to engage their highly agile opponents in a pitched battle in order to bring this stand-off to a successful conclusion. This had the desired effect on the enemy troops who are reported to have been "astonished" at the news of the breaking of the negotiations the previous day.[213] The Greek sources also agree on the main tactic of feigned retreat employed during the first stage of the battle by the Seljuk division in the centre, although the wings of the Turkish army followed a more aggressive stance.[214] This had the desired effect in causing the distance that separated the centre from the flanks of the imperial army to increase, thus breaking its cohesion.

Finally, our sources emphasize the critical stage of the battle that unfolded later that afternoon, when the imperial units approached and captured the now empty Turkish camp, but had not yet succeeded in engaging in a hand-to-hand fighting with the enemy who were retreating but were still within reach. With darkness fast approaching and the Seljuks refusing to offer battle, the emperor naturally feared the possibility of being cut off in enemy territory, hence he ordered his units to retreat back to the (relative) safety of their camp that had been left virtually undefended. The centre of the Byzantine line, commanded by Romanos in person, began to carry out the emperor's order, but the right wing misunderstood this move as a sign of defeat or death of the emperor, and "they welcomed the chance to flee without putting up a fight."[215]

211 Attaleiates, *History*, 285; Skylitzes, *Continuation*, 119.

212 Bryennios, *Ύλη Ιστορίας*, 89. None of the other Greek sources offer any details about the battle array of the imperial army before the battle: Attaleiates, *History*, 287–93; Skylitzes, *Continuation*, 121–23; Zonaras, *Επιτομή*, 215–17.

213 Attaleiates, *History*, 287–93; Skylitzes, *Continuation*, 121–23; Zonaras, *Επιτομή*, 217. Only Bryennios seems to imply that Alp Arslan had calmly prepared his battle plan for Friday's clash: Bryennios, *Ύλη Ιστορίας*, 91.

214 Attaleiates, *History*, 293; Skylitzes, *Continuation*, 121–23; Zonaras, *Επιτομή*, 217; Bryennios, *Ύλη Ιστορίας*, 91.

215 Attaleiates, *History*, 293–95; Skylitzes, *Continuation*, 123; Zonaras, *Επιτομή*, 217; Bryennios, *Ύλη Ιστορίας*, 91–93.

The confusion was exacerbated by the treacherous stand of Andronikos Doukas, who, instead of covering the retreat, began to spread the word that emperor Romanos was now dead and ordered his division to retreat, thus increasing the general confusion. Attaleiates and Skylitzes highlight that this apparent confusion was spotted by Turkish detachments in the foothills of the Süphan Dağl, who immediately sent word to the sultan "urging him to wheel about as he was already in flight and far from his own encampment. Accordingly, he [Alp Arslan] returned at once and the emperor found himself in the thick of the fighting."[216] Romanos tried in vain to recall the troops to his banner, but the centre of the Byzantine line, which he commanded in person, slowly began to retreat, and disintegrate after receiving a devastating barrage of arrows.

It is unfortunate that our Muslim sources devote only a fraction of their narrative to the description of the Battle of Manzikert, and even these do not give us the basic outline of the day's events, let alone any information about the manoeuvres or battle tactics employed by both sides. Al-Ṭurṭūshī, al-Jawzī, al-Athīr, Sibṭ Ibn al-Jawzī and Rashīd al-Dīn make a general reference to the Seljuk attack against the Byzantine army's centre under the leadership of Romanos, with no further comment as to which stage of the battle this attack was launched.[217] Only al-Bundārī and al-ʿAdīm hint that the defeat of the imperial army came as a direct result of the tactics of feigned flight that the Seljuks employed against the Byzantine centre division commanded by the emperor, rather than the treachery of Andronikos Doukas.[218]

216 Attaleiates, *History*, 295–97; Skylitzes, *Continuation*, 123–25; Zonaras, *Επιτομή*, 217; Bryennios, *Ύλη Ιστορίας*, 93.

217 al-Ṭurṭūshī, in Hillenbrand, *Turkish Myth*, 29; al-Jawzī, in Hillenbrand, *Turkish Myth*, 39; al-Athīr, in Hillenbrand, *Turkish Myth*, 65; Sibṭ Ibn al-Jawzī, in Hillenbrand, *Turkish Myth*, 69; Rashīd al-Dīn Faḍlullāh Hamadānī, Hillenbrand, *Turkish Myth*,92.

218 al-Bundārī, in Hillenbrand, *Turkish Myth*, 61; al-ʿAdīm, in Hillenbrand, *Turkish Myth*, 76.

Chapter 3

CHRISTIAN AND MUSLIM SOURCES FOR THE BATTLE OF MANZIKERT

MAKING SENSE OF NUMBERS AND LOCAL TOPOGRAPHY

Military Demography

For military history, the problem of assessing the size and make up of armies is of critical importance, and the question "how big was an army?" echoes in the mind of any historian interested in the history of war. This "million dollar" question has stirred up such a heated debate within the academic community in the last century that historians should tread with caution when referring to issues of military demography.[1] Therefore, it is my intention here to assess the various factors that would have affected the army estimates of the primary sources for the Battle of Manzikert, and to explain to what extent the figures they provide for army sizes are reliable, both in absolute numbers and in the ratios given between mounted and dismounted troops.

In terms of how contemporary historians obtained their numbers, we may consider two broad categories: first, those who had no access to official reports and data because of their position in society and/or because of any secrecy act/policy by the state; and second, those who did have access to official reports and data, and therefore provided numbers based either on military reviews or on regulations or custom. To begin with,

1 As a starting point: Hans Delbrück, *History of the Art of War, Volume II: The Barbarian Invasions (Lincoln: University of Nebraska Press*, 1990), especially 285–316; Delbrück, *Numbers in History* (London: University of London Press, 1913); J. F. Verbruggen, *The Art of Warfare in Western Europe during the Middle Ages* (Woodbridge: Boydell, 1997), 5–9; Verbruggen, "L'armée et la stratégie de Charlemagne," in *Karl der Grosse: Lebenswerk und Nachleben, Vol. 1, Persönlichkeit und Geschichte*, ed. Helmut Beumann (Düsseldorf: Schwann, 1965), 420–36; F.-L. Ganshof, "L'armée sous les Carolingiens," *Settimane di studio del Centro italiano di studi sull'alto medioevo* 15 (1968): 109–30. Delbrück's methods came under criticism by Bernard S. Bachrach, "Early Medieval Military Demography: Some Observations on the Methods of Hans Delbrück," in *The Circle of War in the Middle Ages*, ed. Donald Kagay and L. J. Andrew Villalon (Woodbridge: Boydell, 1999), 3–20. John France, *Victory in the East: A Military History of the First Crusade* (Cambridge: Cambridge University Press, 1999), 122–42. France was also criticised by Bachrach in "The Siege of Antioch: A Study in Military Demography," *War in History* 6 (1999): 127–46. See also: Bachrach, "The Hun Army at the Battle of Chalons (451), An Essay in Military Demography," in *Ethnogenese und Überlieferung. Angewandte Methoden der Frühmittelalterforschung*, ed. Karl Brunner and Brigitte Merta (Vienna: Veröffentlichungen des Instituts für Österreichische Geschichtsforschung, 1994), 59–67. Although Bachrach has vigorously claimed that his opinions on medieval military demography constitute an "'orthodoxy," his views are not widely shared: Bernard S. Bachrach, "Medieval Military Historiography," in *The Routledge Companion to Historiography*, ed. Michael Bentley (London: Routledge, 1997), 203–20; Guy Halsall, *Warfare and Society in the Barbarian West, 450–900* (London: Routledge, 2003), 119–33.

the accounts of historians belonging to the first category would have been affected by three factors, depending on whether they were eyewitness to the events they describe. The first factor relates to the time when their work was written, and the reliance on oral testimonies by the historians. Such accounts always bore the risk of inflation and/or miscalculation. We should also consider that memories inevitably focus on the outcome and significant incidents.[2] In fact, Thucydides observes that battles are difficult enough to reconstruct because each witnesses only knows what is happening in their immediate vicinity.[3] This becomes more acute the lower an eyewitnesses was in the military hierarchy and, as a result, would have had a limited opportunity both to observe and to grasp the reality on the battlefield, which could provide us with a distorted picture of the day's events (although not always).[4] This distorted picture does not apply only to army sizes, the make-up of units, and campaign casualties, but to battle tactics as well, which relates directly to the "command-centred approach" versus the "face of battle approach" to writing a "battle-piece" that we saw in the previous chapter.

Consider, for example, Anna Komnena's reliance on oral testimonies for her campaign and battle narratives that she "gathered first-hand as I often heard the Emperor and George Palaeologus [her mother's brother-in-law and governor of Dyrrachium] talking about them."[5] Although we may have to consider that when Anna refers to actually hearing the account of the events from her uncle, she may have meant a performance of a memoir of events by George that would have written to be recited in the presence of a select audience.[6] It is one thing to record stories of a siege or a battle narrated by your father or uncle who happened to be the commanders *in situ*, regardless of how embellished these stories may be, and another by a mere soldier. In this case, I see no reason to doubt Anna's claim to have recalled stories she had heard in the company of her family. After all, the popularity of the anecdotal episodes in historiographical tradition at this time is also evident in the works of Kekaumenos and Nikephoros Bryennios ("the Younger").[7] However, Anna's insistence that some of her information also came (some decades later) from the eye witness testimony of simple and ordinary men serves

2 The interrogation of participants was a standard element of traditional historical methodology: Polybius, *Histories*, 12: 25; Marincola, *Authority and Tradition*, 63–86; Sinclair, "War Writing," 26–30.

3 Thucydides, *The Peloponnesian War*, bk. 7, chap. 44, 1.

4 On the reliability of eyewitness accounts that provide us with episodes/glimpses of a battle, preserve distorted recollections even of their own contributions, and are unlikely to appreciate broader issues: Michael Whitby, "Reconstructing Ancient Warfare," in *The Cambridge History of Greek and Roman Warfare*, ed. Philip Sabin, Hans Van Wees, and Michael Whitby (Cambridge University Press: Cambridge, 2008), 54–81; Verbruggen, *The Art of Warfare*, 16–18; N. Whatley, "On the Possibility of Reconstructing Marathon and Other Ancient Battles," *Journal of Hellenic Studies* 84 (1964): 119–39 (especially 128–30).

5 Komnena, *Alexiad*, bk. 14, 7. See also: Komnena, *Alexiad*, bk. 1, 6.

6 Neville, *Heroes and Romans*, 48; Sinclair, "War Writing," 98–103.

7 For Bryennius ("the Younger"), Howard-Johnston, "Anna Komnene and the *Alexiad*," 282–88. Also briefly mentioned in Tsougkarakes, *Ύλη Ιστορίας*, 18. For Kekaumenos, see the online introduction in: Kekaumenos, *Consilia et Narrationes*, Greek text, English translation and commentary by

to "shield" her account from accusations of bias and flattery; individuals who provided Anna with information must have been relatives of her father and his leading generals, and therefore hardly insignificant members of the military elite.[8]

Procopius also states his qualifications as a historian by emphasizing that he was not only a first-hand witness, but also an advisor to the most important general of the age, Belisarius.[9] This means that even if Procopius did not witness everything, he certainly had access to those closest to the commander-in-chief, including six key figures in the imperial camp: Peter from Thrace (one of Solomon's spearmen), Marcellus (a *comes excubitorum*), George (one of Belisarius's confidants), Paul of Cilicia (a cavalry officer), Sinnion (a leader of the Kutrigurs), and Ortaias (a ruler of the Moors).[10] Finally, Susan Edgington gives a short but succinct explanation of the strengths and weaknesses in the use of oral sources by Albert of Aachen in his *Historia*. The simple knights who returned to the Rhineland to tell their "stories" inevitably gave their own versions of events in terms of numbers, distances, time, place names, etc., whilst "more insidiously, humans yield to the temptation to make a good story, to show themselves in a flattering light, to exaggerate the hardships, perhaps, and to minimize their own weaknesses."[11] She concludes that the characteristics of oral history "signal caution to the modern reader, but they also account for the *Historia*'s vivid readability."

Second, there is always the possibility (no matter how slim) of historians providing us with inflated (or deflated) numbers according to their biases and sympathies, something akin to modern crowd estimates published in the news or in police reports.[12] For example, George Kennedy has argued that the early Abbasid sources consistently suggest that the Abbasids had many fewer troops than their Umayyad opponents, and this may simply be a device to glorify them and show that God was on their side.[13] N. Whately has noted also that Procopius's numbers in the *Vandal Wars* are much less precise than in the *Persian Wars*, a feature which avoids drawing the readers' attention to the two forces—Roman and Vandal—that were likely similar in size, which would have served to undermine the perceived magnitude of the Roman victory.[14] In fact, there is a remark-

Charlotte Roueché (2013): https://ancientwisdoms.ac.uk/library/kekaumenos-consilia-et-narrationes/intro-kekaumenos/index.html, accessed October 28, 2022.

8 Leonora Neville, *Anna Komnene: The Life and Work of a Medieval Historian* (New York: Oxford University Press, 2016), 83–86; Sinclair, "War Writing," 101–2.

9 *History of the Wars*, bk. 1, chap. 1, para. 3.

10 Conor Whately, *Procopius on Soldiers and Military Institutions in the Sixth-Century Roman Empire* (Leiden: Brill, 2021), 42–45. On Procopius's access to the six figures: Dariusz Brodka, "Prokop von Kaisareia und seine Informanten: ein Identifikationsversuch," *Historia: Zeitschrift für Alte Geschichte* 65 (2016), 108–24.

11 Edgington (ed. and trans.), *Albert of Aachen*, xxvi–xxvii.

12 Bachrach, "Early Medieval Military Demography: Some Observations," 5–8. This is a critique of Delbrück, *Numbers in History*, 12–18. See also: France, *Victory in the East*, 124.

13 Hugh Kennedy, *The Armies of the Caliphs: Military and Society in the Early Islamic State* (London: Routledge, 2001), 97.

14 Conor Whately, "Some Observations on Procopius' Use of Numbers in Descriptions of Combat in

able passage in Ibn-Khaldun's *Muqaddimah* (written in 1377) where the fourteenth-century Arab historian condemns the uncritical use of sources which produce gross exaggerations because of "the common desire for sensationalism…which leads to failure to exercise self-criticism about one's errors and intentions."[15] I will show further down that the Muslim sources for the Battle of Manzikert consistently inflated the numbers of Romanos's army (reporting either six hundred thousand or three hundred thousand) to enhance Alp Arslan's victory.

Third, we should always consider the historian's experience in military matters, as a dismounted knight might be counted as infantry by an inexperienced witness, whereas every adult male was a soldier in steppe armies since no distinction between soldiers and civilians existed. It is for this reason that Polybius argued that personal experience in war and politics was a necessary qualification for historians, even when they derived their information from the questioning of witnesses.[16] For that reason, the *Strategikon* (ca. 600) advised commanders of two things: first, to deploy their troops in varying divisions and spread their men so that the enemy could not simply calculate the size of the army by counting the number of divisional standards;[17] second, if the battlefield was relatively flat and the Roman general could not conceal his numbers from the enemy, then he was to "deceive" his troops by spreading a rumour that the enemy army was mostly made up of baggage trains and spare horses, not actual soldiers.[18] In a well-known example from the First Crusade, on the eve of the Battle of Ascalon (August 12, 1099), the crusaders seized huge herds of animals gathered to feed the Fatimid army defending Ascalon. According to Raymond of Aguilers and Fulcher of Chartres, the crusader army moved with the herds the following morning, suggesting to the enemy that it was far bigger than it actually was.[19]

For those authors who had access to official reports and military rolls,[20] we should bear in mind that the Byzantine army, as a continuator of the Roman army, the Abbasid armies until the 930s, and the Ghaznavid armies all had sophisticated systems of recruitment, review, and remuneration based on the enrolment of soldiers in military registers that, as with any other aspect of their military organization, reflected both the political and economic situation each state found itself in different periods of his-

Wars Books 1–7," *Phoenix* 69 (2015): 394–411 (especially 400–404).

15 Ibn-Khaldun, *The Muqaddimah: An Introduction to History*, ed. Franz Rosenthal and Nessim J. Dawood (Princeton: Princeton University Press, 1967), 11–13.

16 Polybius, *Histories*, 12:27–28.

17 Maurice's *Strategikon: Handbook of Byzantine Military Strategy*, trans. George T. Dennis (Philadelphia: University of Pennsylvania Press, 1984), 102–3. Frontinus also describes several stratagems from antiquity used to create the illusion of superior numbers: Stratagems, *Aqueducts of Rome*, trans. C. E. Bennett and Mary B. McElwain (Cambridge, MA: Harvard University Press, 1925), 2:4:6, 4:7:20.

18 Maurice's *Strategikon*, 71.

19 France, *Victory in the East*, 361.

20 The best analysis of the topic of military dispatches, bulletins, or reports in Byzantium is by Sinclair, "War Writing," chap. 2.

tory.[21] Therefore, we should be fairly certain that in all the aforementioned armies, although we are almost in the dark about the situation with the early Seljuk forces,[22] troops were counted and their numbers were recorded each time by highly skilled clerks. Hence, it should come as no surprise that some of these official numbers found their way into historical accounts of major geopolitical events (battles, sieges, etc.).

To give a handful of examples, John Lydus, who was writing under Justinian, reported that Diocletian had an army of 389,704 and a navy of 45,562, figures that could be considered accurate since John worked in the office of the praetorian prefect. On the other hand, Lydus's near contemporary Agathias detailed that there were "now" one hundred fifty thousand men to defend the whole empire, whereas "previously" there were six hundred forty-five thousand; Agathias was hostile towards Justinian and, as a result, his figures may be exaggerated to show how far the imperial forces had declined in numbers during his reign.[23] As Belisarius's adviser, Procopius would have been able to consult the military rolls, as he refers to the official roll of the African army in his *Vandal Wars*,[24] and in several occasions in his *History* he mentions the establishment and numbers of the armies of Armenia, the East, Africa, and Illyricum, as well as the imperial fleet.[25] Procopius's "job description" could have included the drafting of the military dispatches/reports of the imperial army, although, we cannot be sure. However, he would certainly have had access to letters, reports, and embassies sent between Justinian, his commanders in operational theatres, and other dignitaries.[26] And according to Ian Colvin, archived documents provided the key sources that both Procopius and Agathias used to construct their military narratives, and, as such, the role of oral sources should be diminished. Anna Komnene also writes of commanders such as George Palaeologus, Leo Kephalas, John Doukas, and Eustathios Kamytzes drafting reports on the request of her father about different military situations, to which she refers as *διὰ γραμμάτων*, *διὰ γραφῶν*, or similar forms.[27]

---

21 Constantine Porphyrogenitus, *Three Treatises on Imperial Military Expeditions* , ed. and trans. John F. Haldon (Vienna: Österreichischen Akademie der Wissenschaften, 1990), 84–92; Haldon, *Warfare, State and Society*, chap. 4; Whately, *Procopius on Soldiers*, 48–52; Clifford E. Bosworth, *The Ghaznavids: Their Empire in Afghanistan and Eastern Iran, 994–1040* (New Delhi: Munshiram, 1992), 122–28; Bosworth, "Recruitment, Muster and Review in Medieval Islamic Armies," in *War, Technology and Society in the Middle East*, ed. V. J. Parry and M. E. Yapp (London: Oxford University Press, 1975), 59–77; Kennedy, *Armies of the Caliphs*, chaps. 1–3.

22 Peacock, *Early Seljuq History*, 81–82; Nicolle, *Manzikert* 1071, 27–28.

23 Warren Treadgold, *Byzantium and its Army*, 284–1081 (Stanford: Stanford University Press: 1995), 59–64.

24 *History of the Wars*, bk. 4, chap. 16, para. 3.

25 Treadgold, *Early Byzantine Historians*, 219 (particularly 219n177); Treadgold, *Byzantium and its Army*, 47, 60–64.

26 Ian Colvin, "Reporting Battles and Understanding Campaigns in Procopius and Agathias: Classicising Historians' Use of Archived Documents as Sources," in *War and Warfare in Late Antiquity*, ed. Alexander Sarantis and Neil Christie (Leiden: Brill, 2013), 572–97 (especially 580–92).

27 Komnena, *Alexiad*, bk. 4, i, ii; Sinclair, "War Writing," 193.

I mentioned in a previous chapter that Attaleiates accompanied Emperor Romanos in his campaigns as a *krités tou stratopédou* ("judge of the army"). We may consider this office as akin to a modern military judge and, considering the numerous nationalities that Romanos led against the Turks, special civil judges would have been required "for jurisdictional purposes as *collegia/koina* were subject to the courts of their own officers or commanders."[28] Therefore, resolving issues of conflicting jurisdiction would have been the main duty of Attaleiates, as the frustrated historian moans in his work that "I am busy with thousands of responsibilities in the army; am continually whirled about by legal disputes, and spend hours working away on the documents that result from them."[29] Yet, although we know that Attaleiates was consulted several times about strategy, he certainly was not always privy to Romanos's decision-making. On top of that, there were several high-ranking officials from the imperial court who accompanied Romanos in 1071, men like "Leon, who was *epi ton deeseon*...; also, the *magistros* and *protasekretis* Eustratios Choirosphaktes...the *protovestes* Basileios Maleses, the emperor's closest associate,"[30] but we cannot be sure about the access Attaleiates had to them. Nor can we be sure about his access to the official reports and military rolls of the campaign,[31] although it has been suggested that Attaleiates's knowledge of Romanos's heroic actions in the field and subsequent time in captivity was taken from a letter Diogenes dispatched to his wife.[32]

There is also a very interesting discussion on the numbers provided by Ammianus Marcellinus, a member of the elite *protectores domestici* (a sort of an adjutant or staff officer),[33] for Emperor Julian's expedition against Sassanid Persia in 363, especially when compared to Zosimus's figures for the same expedition. For example, Jan den Boeft et al have argued that there is polemic behind the discrepancy between Ammianus's figure of two thousand five hundred (24:2:22) for the Persian prisoners taken at Pirisabora city on the Euphrates, and the five thousand claimed in Zosimus (3:18:5).[34] Finally, because every military operation certainly entailed a mass of administrative records, John F. Haldon's investigation into the documents about the Byzantine expeditions in

---

28 Haldon, "The Krites Tou Stratopedou," 280.

29 Attaleiates, *History*, 11.

30 Attaleiates, *History*, 303.

31 Pérez Martín has suggested that Attaleiates may have accessed documentary sources through his office: Pérez Martín, *Historia*, xliii and 127. Sinclair wraps up his section on Attaleiates with the following: "Not since Ammianus Marcellinus and Procopius had a Roman historian enjoyed such close proximity with central command." "War Writing," 89.

32 For more on that alleged letter: Sinclair, "War Writing," 87–88 and n450.

33 The military duties of the protectores domestici included participation in acts of espionage or covert operations, the gathering of intelligence, and participation in the planning of military operations akin to modern-day staff officers. Kagan, *The Eye of Command*, 25–26.

34 Jan den Boeft, Jan Willem Drijvers, Daniël den Hengst, and Hans C. Teitler, *Philological and Historical Commentary on Ammianus Marcellinus XXIV* (Leiden: Brill, 2001), 66. See the more detailed discussion in: Neil McLynn, "The Persian Expedition," in *A Companion to Julian the Apostate*, ed. Stefan Rebenich and Hans-Ulrich Wiemer (Leiden: Brill, 2020), 293–325; Charles W. Fornara, "Julian's Persian Expedition in Ammianus and Zosimus," *Journal of Hellenic Studies* 111 (1991):

Crete (911 and 949) and in southern Italy (934 and 935), found in chapters forty-four and forty-five of Constantine VII's *De Ceremoniis*,[35] suggests that the Cretan and Italian "inventory lists" of men, resources, and the expenditure involved are certainly not as detailed as the original documents, but rather abridged versions pieced together from the archives of several governmental departments in the capital.[36] Haldon has further suggested that the Italian "inventory list" was updated following the end of the mission which, one could argue, proves the central government's interest in documenting the consequences and results of major operations.[37] Therefore, the key issue is what kind of access each historian had to these official reports and data.

These official numbers, set by regulations or custom, should have corresponded fairly closely to actual sizes except in unusual circumstances. For example, not all military rolls would have reflected all the deaths, retirements, desertions, or the number of soldiers who would have been unable to serve for financial reasons;[38] pre-battle and/or annual reviews would have helped keep the rolls updated.[39] However, numbers could still be rounded off and, as a result, some modern critics may feel that all rounded numbers mentioned in medieval sources are likely to be wildly exaggerated, or at least formulaic and unreliable. Yet, some historians have argued that numbers for official establishments can be round without being either rounded or estimated, as the overall principle of standardization in the armies of many centralized states is indisputable.[40] And it is easy to imagine that efficient record-keeping and standardization in the armies would

1–15. On the verisimilitude of Ammianus's numbers for the Battle of Strasburg for which he was not an eyewitness but, nevertheless, had plentiful reliable sources: Kagan, *The Eye of Command*, 61, 63–65.

35 *Constantine Porphyrogennetos: The Book of Ceremonies*, ed. and trans. Ann Moffatt and Maxeme Tall (Leiden: Brill, 2012), 651–78.

36 John F. Haldon, "Theory and Practice Tenth-century Military Administration: Chapters II, 44 and 45 of the Book of Ceremonies," *Travaux et mémoires* 13 (2000): 201–352. See also: Warren Treadgold, "The Army in the Works of Constantine Porphyrogenitus," *Rivista di Studi Bizantini e Neoellenici* 29 (1992): 100–40. A valorous attempt to build a comprehensive picture of the numbers involved in Byzantium's Cretan campaign of 960–61: Lucas McMahon, "Logistical Modelling of a Sea-Borne Expedition in the Mediterranean: The Case of the Byzantine Invasion of Crete in AD 960," *Mediterranean Historical Review* 36 (2021): 63–94.

37 Haldon, "Theory and Practice," 257; Sinclair, "War Writing," 196.

38 On the discrepancy between "paper strengths" and "actual strengths" in Roman and medieval armies: Carl Sverdrup, "Numbers in Mongol Warfare," *Journal of Medieval Military History* 8 (2010): 109–17 (especially 111); Brian Campbell, *War and Society in Imperial Rome, 31 BC–AD 284* (London: Routledge, 2002), 27–28; Halsall, *Warfare and Society*, 125; Haldon, *Warfare, State and Society*, 102.

39 On army reviews: Tilemachos Loungis, "Επιθεώρηση ενόπλων δυνάμεων πριν από εκστρατεία [The Review of Armed Forces before Campaign], in *Byzantium at War (9th–12th Century)*, ed. K. Tsiknakes, 93–110 (Athens: Institute for Historical Research, 1997); Kennedy, *Armies of the Caliphs*, 21–23, 72–73, 157; Bosworth, *The Ghaznavids*, 122–26; Nicolle, *Crusader Warfare*, 2:56–57.

40 Warren Treadgold, "Standardized Numbers in the Byzantine Army," *War in History* 12 (2005): 1–14; Treadgold, *Byzantium and its Army*, 43–86; Kennedy, *The Armies of the Caliphs*, xii–xiii, 19–21.

have proved useful to the campaign planners when calculating the logistical needs of contemporary campaigns; after all, enemies change, but distances do not.[41]

An examination of Byzantine primary sources in terms of the frequency and accuracy of the numbers they provide for Romanos's campaigns between 1068 and 1071 would leave the reader frustrated and disappointed, considering what I just mentioned about the possible access to official data from military reviews, and the *kodikes* (registers) that were held in each theme and in the central government department of the military *logothesion* by officials such as the *chartoularioi* and the *protonotarioi*.[42] Hence, the army estimates we read in the account of Skylitzes are vague, and are therefore of little use. For example, he refers to the Turkish parties that raided eastern Anatolia in 1068 simply as "irresistible," while the Byzantine forces that responded to that threat were (in all honesty) "by no means of the sort worthy of a Roman emperor but such as the times provided." Yet, Skylitzes goes on to enumerate these units "of Macedonians and Bulgarians, of Kappadokians and Uzes and foreign peoples who for some reason happened to be on hand, as well as Franks and Varangians."[43] The same is also true about the details on Romanos's campaign of 1069, when the emperor is described as bringing to Kaisarea "a large force" after "he learned that a great many Turks were pillaging the area."[44] Sadly, there is not even a hint about the numbers involved in the 1071 campaign, although Skylitzes again indicates the presence of contingents "of Uze mercenaries and the Franks under Rouselios."[45]

Equally frustrating and disappointing is the paucity of numbers for the imperial campaigns of 1068–1071 in the accounts of the rest of the Byzantine sources. Bryennios provides a sole comment about the "few Roman casualties" after Manuel Komnenos's defeat and capture by Alp Arslan's brother-in-law, Arisghi, in the summer of 1070 that led to an alliance between the two commanders.[46] Zonaras's only estimate of an army size comes for Romanos's campaign of 1068, for which he adds the negligible comment that the emperor left the capital "without a sizeable army."[47] Then, there is Attaleiates's inexact comments about the "good-sized force"[48] that was dispatched against a devastating Turkish raid on the territory of Melitene and Kaisarea in the summer of 1067, cou-

41 "We have no precise knowledge of how long older records and plans were kept, but some knowledge survived of aborted as well as unsuccessful military expeditions and invasions." Walter E. Kaegi, "Byzantine Logistics: Problems and Perspectives," in *Feeding Mars: Logistics in Western Warfare from the Middle Ages to the Present*, ed. John A. Lynn (Boulder: Westview, 1993), 39–55 (quotation at 47).

42 For the chartoularioi: Kazdhan, ed., *Oxford Dictionary of Byzantium*, 1:416. For the protonotarioi: Kazdhan, *Oxford Dictionary of Byzantium*, 3:1746.

43 Skylitzes, *Continuation*, 83.

44 Skylitzes, *Continuation*, 99.

45 Skylitzes, *Continuation*, 115.

46 Bryennios, *Ύλη Ιστορίας*, 75.

47 Zonaras, *Επιτομή*, 203.

48 Attaleiates, *History*, 175.

pled with the equally vague reference to the "fairly large contingent,"[49] which included Frankish mercenaries left by the emperor in Melitene to watch over the frontier districts and fend off Afshin's forces while he continued his march south to Syria. Attaleiates's account of Romanos's campaign of 1068 includes some important details about the main units that formed the core of the imperial army, namely from the "west [i.e., from the Balkans] and Kappadokia [Romanos's homeland], and [he] sent an invitation to the Skythians [i.e., Patzinaks and/or other Turkic groups]."[50] Further down, he also includes a reference to the participation of (eastern) tagmata,[51] adding that "It was something to see the famous units and their commanders now composed of just a few men," and of "warlike and bloodthirsty" Frankish and Armenian mercenaries.[52] Attaleiates also refers to the Turkish raiding parties that were invading the area between Melitene and the northernmost branch of the Euphrates Valley in 1069 to be "in greater numbers than ever before,"[53] a comment that is equally vague as the one about the "sufficient force"[54] of Nikephoros Bryennios the "Elder" that was dispatched by the emperor (on August 24, 1071) to support the foragers of the imperial army at Manzikert, and was eventually ambushed and retreated towards the Byzantine camp.

Crossing over to the Armenian accounts for the Battle of Manzikert, one would expect Matthew of Edessa's *Chronicle* to include at least some estimates for the Turkish armies that were penetrating eastern Asia Minor between 1068 and 1071, since the second part of the work covering the period between 1051 and 1101 largely depends on contemporary eyewitness accounts.[55] Nonetheless, Matthew's account is surprisingly vague and uninformative when it comes to numbers; for example, he remarks that Romanos "collected a great number of troops" for his first campaign in 1068, adding further down that Alp Arslan conquered Artchesh and Manzikert, in August/September 1070, "with a tremendous number of troops."[56] Matthew further refers to Romanos's army in 1071, comprising "very great and formidable number of troops"[57] from many different regions of the empire, while the sole number we get for this critical campaign concerns Trachaniotes's force hurried to Khliat, which comprised some thirty thousand men. Matthew is the only source that reports an alleged dispatch of another twelve

---

49 Attaleiates, *History*, 197.

50 Attaleiates, *History*, 189.

51 Haldon believes that Attaleiates uses the term tagmata to refer to the units drawn from the regular military registers of each thema, a view that is also shared by the editors of *The History of Michael Attaleiates* who translate the term tagma as "unit." Haldon, "Approaches," 48; Attaleiates, *History*, 188–89.

52 Attaleiates, *History*, 197–99.

53 Attaleiates, *History*, 229.

54 Attaleiates, *History*, 281.

55 Dostourian, ed., *Armenia and the Crusades*, 1. For Matthew's sources for the *Chronicle*: Andrews, *Matt'eos Urhayec'i*, 15–22.

56 Matthew of Edessa, *Chronicle*, 128, 130.

57 Matthew of Edessa, *Chronicle*, 132.

thousand troops to the "Abkhazes."[58] Summarizing Matthew of Edessa's *Chronicle* for the period up to 1129, the *Chronicle* of Smbat Sparapet includes the thirty thousand soldiers sent to Khliat under Trachaniotes, although Smbat adds that only ten thousand instead of Matthew's twelve thousand were dispatched to Abkhazia.[59]

Surprisingly, Muslim sources offer more army estimates about Romanos's campaigns than their Chalcedonian and non-Chalcedonian counterparts. Both al-Ṭurṭūshī and al-Qalānisī, our main sources from the twelfth century, put the figure for Romanos's army in 1071 at six hundred thousand men, a formidable force that was "accompanied by an innumerable amount of animals, weapons and mangonels and pieces of equipment made ready for conquering citadels in war."[60] This is, clearly, a wildly exaggerated figure that builds on the great discrepancy in the respective numbers of the opposing armies, especially considering that al-Ṭurṭūshī gives the number for Alp Arslan's troops to as low as twleve thousand.[61] This discrepancy in numbers would have served two purposes: first, it is a great way for the historian to heighten the tension, give a dramatic effect in his narrative, and convey to the reader the scale of the encounter;[62] but, second, it also plays a key role in the glorification of Alp Arslan's victory at Manzikert, one that bore great similarities in every Muslim audience's mind with the success of Muhammad's own triumph against all odds over the pagan Meccans at Badr.[63] Later Muslim sources also document Romanos's army as being overwhelmingly numerous, as high as three hundred thousand, an estimate that remains more or less the same up to Mīrkhwānd's mid-fifteenth-century account.[64] Those for the size of Alp Arslan's army also vary to between twelve-thousand and twenty-thousand strong,[65] although it is only

---

58 Matthew of Edessa, *Chronicle*, 134. A dubious statement by Matthew of an alleged dispatch of military reinforcements to Bagrat IV (King of Georgia, 1027–1072), on the eve of Manzikert. For King Bagrat IV and the Kingdom of Georgia in his time: Ronald Grigor Suny, *The Making of the Georgian Nation* (Bloomington: Indiana University Press, 1994), 33–34.

59 Smbat Sparapet, *Chronicle*, 37.

60 al-Ṭurṭūshī, in Hillenbrand, *Turkish Myth*, 27; al-Qalānisī, in Hillenbrand, *Turkish Myth*, 30.

61 al-Ṭurṭūshī, in Hillenbrand, *Turkish Myth*, 28.

62 There is a great in-depth analysis of Procopius's use of the term myriad, or some other cognate, and of the impact, if any, it had on his descriptions of combat, especially in his Gothic Wars. Whately, *Combat, Culture, and Didacticism*, 171–77.

63 Hillenbrand, *Turkish Myth*, 115.

64 al-Fāriqī, in Hillenbrand, *Turkish Myth*, 34; Nishapuri, in Hillenbrand, *Turkish Myth*, 36; al-Jawzī, in Hillenbrand, *Turkish Myth*, 38 (with a breakdown of the units comprising the army); al-Ḥusaynī, in Hillenbrand, *Turkish Myth*, 53; al-Bundārī, in Hillenbrand, *Turkish Myth*, 59 (with a breakdown of the nationalities comprising the army); al-Athīr, in Hillenbrand, *Turkish Myth*, 64 (two hundred thousand, with a breakdown of the nationalities comprising the army); Sibṭ Ibn al-Jawzī, 68 (over four hundred thousand men and five hundred thousand officers); al-ʿAdīm, in Hillenbrand, *Turkish Myth*, 74 (with a breakdown of the nationalities comprising the army); Rashīd al-Dīn Faḍlullāh Hamadānī, in Hillenbrand, *Turkish Myth*, 90; Mīrkhwānd, in Hillenbrand, *Turkish Myth*, 96.

65 al-Qalānisī, in Hillenbrand, *Turkish Myth*, 30; al-Fāriqī, in Hillenbrand, *Turkish Myth*, 34 ("only small numbers"); Nishapuri, in Hillenbrand, *Turkish Myth*, 36; al-Jawzī, in Hillenbrand, *Turkish Myth*, 38; al-Bundārī, in Hillenbrand, *Turkish Myth*, 58, 60; al-Athīr, in Hillenbrand, *Turkish Myth*,

al-Jawzī and Sibṭ Ibn al-Jawzī who report the size of the sultan's *askar* numbering four thousand *ghulams*.[66] Once again, this discrepancy in numbers aims to add to the myth that developed around the victory of Alp Arslan at Manzikert, as we can also see in the sultan's pre-battle speech to his troops reported by al-Jawzī: "We are with reduced numbers of men. I want to throw myself at them [the Byzantines] when prayers are being said for us and for the Muslims on the pulpits. I will either achieve my aim or I will go as a martyr to Paradise. So, he amongst you who wants to follow me let him follow me and he who wants to leave let him go."[67]

This literary motif and the idea of triumphing against all odds is a typical rhetorical strategy that was also used by the chroniclers who wrote their epic verses about the Norman expansion in the Mediterranean in the eleventh century, showing as they did the Normans's innate *virtus* and *strenuitas*, and that they succeeded through divine favour.[68] In an effort to portray his Norman heroes as "model warriors," in a similar way as the Muslim sources portray Alp Arslan as the "model jihad warrior," Geoffrey Malaterra, the chronicler who promoted the introduction of the *clichés* about the Norman prowess in war,[69] wrote about Duke Robert Guiscard's expedition in the Balkans in 1081[70]: "For the duke was extraordinarily bold and determined in military matters…He dared to undertake a war with only a small force against so populous an empire and an emperor so well provided with troops, against so many thousands of enemies, and he expected to defeat them." Further down, Guiscard gave a rousing speech to his troops after disembarking from his ships on the coast of Epirus, in the autumn of 1081: "Bravest of knights, and not unworthy heirs of the reputation of your predecessors; for this is an inheritance equal to you, though to acquire it you will have to resort to arms! Do not be frightened by a vile and feeble mob, however numerous it may be, or the noise of a horde of enemies; but take encouragement from the appearance and fertility of the enemy's land, which you will acquire as your own hereditary possessions." Perhaps the best (and maddest) example of the creation of a medieval hero comes from Malaterra's description of the

---

64; Sibṭ Ibn al-Jawzī, in Hillenbrand, *Turkish Myth*, 68; al-ʿAdīm, in Hillenbrand, *Turkish Myth*, 74; Rashīd al-Dīn Faḍlullāh Hamadānī in Hillenbrand, *Turkish Myth*, 90; Mīrkhwānd, in Hillenbrand, *Turkish Myth*, 96.

66 al-Jawzī, in Hillenbrand, *Turkish Myth*, 37; Sibṭ Ibn al-Jawzī, in Hillenbrand, *Turkish Myth*, 67.

67 al-Jawzī, in Hillenbrand, *Turkish Myth*, 38–39.

68 Matthew Bennett, "Norman Conquests: Nature, Nurture, Normanitas," in *The Normans in the Mediterranean*, ed. Emily A. Winkler and Liam Fitzgerald (Turnhout: Brepols, 2021), 43–65 (esp. 54–65); Eleni Tounta, "The Norman Conquerors between Epos and Chanson de Geste: The Perception of Identities in Cultural Flows," in *Norman Tradition and Transcultural Heritage Exchange of Cultures in the 'Norman" Peripheries of Medieval Europe*, ed. Stefan Burkhardt and Thomas Foerster (Routledge: London, 2013), 125–47; Graham A. Loud, "The 'Gens Normannorum': Myth or Reality?," in *Conquerors and Churchmen in Norman Italy*, ed. Graham A. Loud (Aldershot: Ashgate, 1999), 104–16.

69 Ovidio Capitani, "Specific Motivations and Continuing Themes in the Norman Chronicles of Southern Italy in the Eleventh and Twelfth Centuries," in *The Normans in Sicily and Southern Italy*, Lincey Lectures 1974 (London: Oxford University Press, 1977), 6.

70 Malaterra, *Deeds*, bk. 3, 24.

Battle of Cerami, which was fought in June 1063 and was one of the most significant battles in the Norman conquest of Sicily (1060–1091), when Count Roger's young and audacious nephew Serlo: "entered the [enemy] camp, not waiting for his uncle's arrival within the walls, bursting through the gates like a raging lion and inflicting great casualties on them [Muslims]. Even though they numbered three thousand, in addition to the infantry—of whom there were an infinite number—wonderful to say, he put them to flight with his thirty-six knights."[71]

Building a portrait of her father as a great warrior and a strategist, Anna Komnena also writes about Alexios's performance at the Battle of Kalavrye (1078),[72] a victory that not only solidified his military credentials, but one that he fought while outnumbered 2:1 by the army of Nikephoros Bryennios "the Elder": "Bryennius, in addition to his confidence in his forces, was protected by their experience and orderliness, whereas Alexios, on the other hand, centered but few, and those very meagre, hopes on his army, but as counter-defence, could rely on the strength of his scientific knowledge and his strategic devices."[73] Anna's effort to mythologize the already heroized Bryennios in her *Alexiad* by likening him to Ares certainly increased the appeal of Alexios and of his victory to her audience as that of another David against a Goliath.[74]

Military Geography and Topography

Crossing over to the value of military geography for the study of historic campaigns and battles, the fact that terrain[75] underpins military engagement has been recognized for millennia. One of the earliest Greek authors of treatises on the art of war, Aeneas (known as "Taktikos"), remarked in his (mid-fourth century BC) work on *How to Survive under Siege* that: "You [general] must always, in making your attacks upon the enemy, strive to profit from your acquaintance with the terrain; for you will have a great advantage from previous knowledge of the country and by leading the enemy into such places as you may wish, which are known to you and suitable, whether for defence, or pursuit, or flight, or withdrawal into the city either secretly or openly."[76] Vegetius further noted in his (ca. 400 AD) *De Re Militari* that "The good general should know that a large part of a victory depends on the actual place in which the battle is fought. Be at pains there-

---

71 Malaterra, *Deeds*, bk. 2, 33.

72 John Haldon, *The Byzantine Wars* (Stroud: Tempus, 2001), 181–87.

73 Komnena, *Alexiad*, bk. 1, 5.

74 Penelope Buckley, *The* Alexiad *of Anna Komnene: Artistic Strategy in the Making of a Myth* (Cambridge: Cambridge University Press, 2014), 56–61.

75 Terrain is defined as "an area of land, when considering its natural features." On the other hand, terrain analysis is "The collection, analysis, evaluation, and interpretation of geographic information on the natural and manmade features of the terrain, combined with other relevant factors, to predict the effect of the terrain on military operations." Department of Defense, *Dictionary of Military and Associated Terms* (November 8, 2020), 367.

76 Aeneas Tacticus, *Asclepiodotus, Onasander, Aeneas Tacticus, Asclepiodotus, and Onasander*, trans. Illinois Greek Club (Cambridge, MA: Harvard University Press, 1928), 16:19.

fore when you are going to engage in combat, to get help first from the place."[77] Five centuries later, we read in Emperor Leo VI's *Taktika* that: "The condition of the terrain has often proved greatly advantageous to the men engaged in combat. There have been times when it enabled the weaker side to become the stronger. For this reason, you must accommodate your battle plans to such terrain in which you know that your army has an advantage and the enemy are at a loss."[78]

Because in any military action there are two basic levels of engagement, the strategic, "by which a commander may defend his own lands and defeat his enemies," and the tactical, "which enables one to organize and manoeuvre a body of armed men in an orderly manner."[79] Here I will assess the degree of familiarity of our primary sources with the areas central to the Byzantine campaigns between 1068 and 1071 against the Seljuks, and I will ask how accurate their description of battlefields and army campaign routes are.[80]

The *History* of Michael Attaleiates and *The Continuation of the Chronicle of John Skylitzes* are fundamental for the reconstruction of the strategy and routes of Romanos's armies between 1068 and 1071. Both accounts point out the shift in focus of the main Seljuk attacks in 1068, from the Armenian highlands and the Araxes and Arsanias River valleys to Edessa, Aleppo, and the region of Cilicia; apparently, this is how we should explain Romanos's advance southeast to Lykandos, to the west of Melitene, instead of northeast to Sebasteia and Koloneia, in preparation for an invasion of Syria after the scorching summer-months of 1068.[81] This is confirmed by Romanos's strategic reaction (his march back north to Sebasteia) to the "utterly unexpected"[82] Turkish attack and pillage of Neokaisarea on the Lykos River, around 160 km to the northwest of Koloneia. In fact, Attaleiates calls Romanos's eight-day pursuit of the Turkish band across to Tephrike (modern Divriği) "a truly admirable and memorable" feat.[83]

Regarding the sequence of events in the main operational theatre of Syria, Attaleiates and Skylitzes report that on October 4, 1068 the Byzantine forces set off from Sebasteia via Koukousos (modern Göksun), Germanikeia (modern Kahramanmaraş),

77 Vegetius, *De Re Militari*, bk. 3, 13.

78 *The Taktika of Leo VI*, cons. 20, 56.

79 "The Anonymous Byzantine Treatise on Strategy," in *Three Byzantine Military Treatises*, ed. and trans. George T. Dennis (Washington, DC: Dumbarton Oaks, 2008), §§4–5, §14.

80 On the importance of terrain in Military History: Verbruggen, *Art of Warfare*, 7–9; Russel S. Harmon, Francis H. Dillon III, and John B. Garver Jr., "Perspectives on Military Geography," in *Studies in Military Geography and Geology*, ed. Douglas R. Caldwell, Judy Ehlen, and Russell S. Harmon (London: Kluwer, 2004), 7–20; Peter Doyle and Matthew R. Bennett, "Terrain in Military History: An Introduction," in *Fields of Battle, Terrain in Military History*, ed. Peter Doyle and Matthew R. Bennett (London: Springer, 2002), 1–7; in the same volume: Kerry Cathers, "'Markings on the Land" and Early Medieval Warfare in the British Isles," 9–17.

81 Attaleiates, *History*, 193; Skylitzes, *Continuation*, 83; Beihammer, *Byzantium and the Emergence of Muslim-Turkish Anatolia*, 138.

82 Attaleiates, *History*, 193; Skylitzes, *Continuation*, 87; Zonaras, *Επιτομή Ιστοριών*, 3:204–5.

83 Attaleiates, *History*, 195; Skylitzes, *Continuation*, 87.

and the theme of Teluch (modern Dülük) to Aleppo where, following the pillaging of the rural environs of the city, Romanos marched for three days to Hierapolis (modern Manbij), some 90 km northeast of Aleppo.[84] Both primary sources include a rare topographical description of the plains around Hierapolis, where they emphasize the flatness of the ground that made it ideal for horse-riding, while also noting the irrigation canals to the west of the city that posed an obvious obstacle for cavalry units.[85]

Following the establishment of an imperial garrison at Hierapolis, Attaleiates and Skylitzes write that the imperial army marched west to Azaz, some 100 km west of Hierapolis and around 50 km north of Aleppo, where it failed to take the fortress-town not only because, as both emphasize, of its strong double walls and its location on the ridge of a hill, but also because of the lack of water supplies in the region.[86] Hence, Romanos and his troops continued their march in a westerly direction, burning down the village of Katma (modern Qatma), just 7 km west of Azaz, and defeating yet another Muslim attack at a place named Terchala, probably on the eastern bank of the Afrin River (modern Nahr 'Ifrīn), before driving out the Arab garrison from the fortress of Artah (modern-day Reyhanli), around 65 km southwest of Azaz.[87] Our sources report that the imperial forces did not spend the winter in Antioch, because the city's countryside had been recently ravaged by Arab forces, heading home instead via the *kleisoura* of Podandos, in the southernmost part of Kappadokia, in late December (1068).[88]

The failure of the imperial commander stationed in Melitene to relieve Amorion, which was being besieged and was eventually sacked by the Turks in the autumn of 1068,[89] a military setback that left the emperor "distressed, but unable to take any further action" because of the onset of winter, gave rise to a significant increase of hostile activities in the Upper Euphrates and the Anti-Taurus regions for 1069. Romanos had clearly lost the strategic initiative and, as Attaleiates reports, he set out once more from Kaisarea to Larissa (modern Mancilik) in the theme of Sebasteia, around 180 km northeast of Kaisarea, to try to "plug the gap" in the imperial defences by chasing away a strong Turkish force all the way to Melitene.[90]

Attaleiates is critical of Romanos's decision not to march to Khliat by Lake Van, apparently changing his mind at the last minute before the city of Romanopolis (modern Bingöl) and instead choosing to march northwest across the Mouzouron Mountains (modern Mezūr Dāg) to the small theme of Keltzine (modern Erzincan) with only a fraction of his army.[91] This proved to be a strategic blunder that allowed the Turks roaming

84 Attaleiates, *History*, 197–99; Skylitzes, *Continuation*, 87–89; Zonaras, *Επιτομή Ιστοριών*, 3:204–5.

85 Attaleiates, *History*, 203–5; Skylitzes, *Continuation*, 91.

86 Attaleiates, *History*, 213–15; Skylitzes, *Continuation*, 93–95.

87 Attaleiates, *History*, 215–17; Skylitzes, *Continuation*, 95.

88 Attaleiates, *History*, 217–21; Skylitzes, *Continuation*, 95.

89 Attaleiates, *History*, 223; Skylitzes, *Continuation*, 95.

90 Attaleiates, *History*, 229–33; Zonaras, *Επιτομή Ιστοριών*, 3:208–9.

91 Attaleiates, *History*, 241–43; Skylitzes, *Continuation*, 101. In fact, Attaleiates's inability to understand (κατόπιν ἀκληρία: following misfortune) Romanos's operational decision may have

the region to inflict a crushing defeat on the imperial units that were left in Romanopolis under Philaretos Brachamios, before penetrating deep into Kappadokia as far as Ikonion, which they looted.[92] Romanos waited at Keltzine to receive the injured men from Brachamios's units, fearful of possible Armenian attacks on his army, a clear sign that the emperor no longer trusted the local Armenian populations of the region. Certainly, it was Attaleiates who intervened again in the strategic decision-making process and convinced Romanos of how "unprofitable" (*ἀσύμφορον*) a march against Melitene would be, suggesting, instead, they should concentrate their efforts on the defence of Kappadokia, the defences of which had not yet completely disintegrated.[93] Yet, Romanos's forced march to Heraklea (modern Ereğli) at the "Cilician Gates" could not prevent the escape of the looters of Ikonion to Syrian territory, although they were attacked in Seleukia (modern Silifke), some 190 km south of Heraklea, by some local Armenian soldiers.[94]

The accounts of both Attaleiates and Skylitzes are significantly shorter for the defensive operations led by the *protoproedros* and *kouropalates* Manuel Komnenos in 1070. Nonetheless, both notice that Manuel assumed important military and administrative powers to operate in the region of Kappadokia, with his headquarters established in Kaisarea.[95] However the critical move was the weakening of Manuel's army by the emperor following the dispatching of a significant part of Manuel's army south to relieve the besieged population of (the Syrian) Hierapolis. Finally, it is on the occasion of the intra-dynastic conflict within the Seljuk ruling house following the revolt of Alp Arslan's brother-in-law Arisghi, that Attaleiates and Skylitzes report a band of Turkish invaders reaching the westernmost point of their incursions in Asia Minor, and pillaging the city of Chonai (modern Honaz) in the Upper Meander Valley, "news that greatly depressed us, for it was as though these disasters were being caused by divine anger."[96]

On March 13, 1071, the feast of Orthodoxy, Emperor Romanos crossed the Bosphorus for his last expedition in the East. The imperial army followed the "northern route" leading to the theme of Anatolikon via Dorylaion, across the Sangarios River at the Bridge of Zompou, and across the Halys River into the theme of Charsianon. Attaleiates notes that the emperor and his army passed by the vicinity of Kaisarea but did not allow the bulk of his troops to enter the city, probably deciding to camp in the vicinity of the city in order to spare it the burden of having to accommodate thousands of troops, especially considering that Kaisarea had served as the base for the 1070 campaign under Manuel Komnenos.

---

been because the historian was posted in a different section of the army at the time the emperor made his final decision: Vratimos-Chatzopoulos, "Two Expeditions," 223–24.

92 Attaleiates, *History*, 245–47; Skylitzes, *Continuation*, 101–3; Zonaras, *Επιτομή Ιστοριών*, 3:208–9.

93 Attaleiates, *History*, 247–51.

94 Attaleiates, *History*, 251; Skylitzes, *Continuation*, 105. Bryennios notes that Manuel was based in the theme of Khaldea: *Ύλη Ιστορίας*, 72–73.

95 Attaleiates, *History*, 253–55; Skylitzes, *Continuation*, 107.

96 Attaleiates, *History*, 257; Skylitzes, *Continuation*, 107–9; Zonaras, *Επιτομή Ιστοριών*, 3:210–11.

Attaleiates then mentions that the emperor set up camp for many days at a place called Krya Pege ("Cold Spring"), for which Michael provides topographical explanation for the origins of its name. Then, the army marched to Sebasteia and reached Koloneia and later Theodosiopolis.[97] Previously, Alp Arslan had conquered Manzikert in the autumn of 1070 and, appreciating its strategic importance, had a garrison of Turks and "Dilimnitai" installed there. Our sources emphasize that the incessant fighting and the devastation inflicted in the area between Theodosiopolis and Manzikert forced Romanos to order his troops to carry provisions for two months after leaving the relative safety of Theodosiopolis.[98] The strategic target this time would be the city of Khliat,[99] following Attaleiates's advice to Romanos at the war-council two years before, apparently in an effort to recover the important fortresses to the north of Lake Van that controlled the main invasion route north of the Anti-Taurus Mountains that led into central Anatolia. Sadly, however, both Attaleiates's and Skylitzes's accounts lack any meaningful topographical details about the battlefield of Manzikert.

Regrettably, Muslim sources are much less informative about the geography and topography of the areas where the events they describe took place. Information about the topography of the Manzikert battlefield is negligible, as only al-Qalānisī simply writes of "a valley there, where the two sides had met, was filled [with corpses]."[100] The location of the battle is named several times as a plateau known as *al-Rahwa* or *al-Zuhra*,[101] yet there is still some controversy over its exact location. Nicolle and Leveniotes appear convinced that this plateau stretched south and southeast of Manzikert, which makes sense considering that al-Ḥusaynī, al-Bundārī, and al-ʿAdīm note that *al-Rahwa* (or *al-Zuhra*) was indeed somewhere between Khliat and Manzikert.[102] However, a recent reassessment of archaeological and textual evidence on Urartian routes used for military campaigns reconsidered the location of the Rahwa (or Rahva) Plateau argues that it was located along the western route departing from Tatvan that connected Lake Van to the Elaziğ-Malatya route in the Muş Plain.[103] A 1987 archaeological survey also places the Rahwa Plateau on the south slopes of Mount Nemrut (Nemrut Daği), on the western side of Lake Van.[104]

---

97 Attaleiates, *History*, 265–69; Skylitzes, *Continuation*, 113–15; Bryennios, *Ύλη Ιστορίας*, 74–81; Zonaras, *Επιτομή Ιστοριών*, 3:212–13; Psellos, *The History*, VII. 19.

98 The area is described as "ἀοίκητον" and "ἠρημωμένη": Attaleiates, *History*, 271; Skylitzes, *Continuation*, 115; Zonaras, *Επιτομή Ιστοριών*, 3:212–13.

99 Attaleiates, *History*, 271; Skylitzes, *Continuation*, 115; Zonaras, *Επιτομή Ιστοριών*, 3:212–13.

100 al-Qalānisī, in Hillenbrand, *Turkish Myth*, 30.

101 al-Jawzī, in Hillenbrand, *Turkish Myth*, 38; al-Ḥusaynī, in Hillenbrand, *Turkish Myth*, 54; al-Bundārī, in Hillenbrand, *Turkish Myth*, 60; Sibṭ Ibn al-Jawzī, in Hillenbrand, *Turkish Myth*, 68; al-ʿAdīm, in Hillenbrand, *Turkish Myth*, 75.

102 Nicolle, *Manzikert*, 62; Leveniotes, *Η πολιτική κατάρρευση του Βυζαντίου*, 170n903.

103 Ali Çifçi and Bilcan Gökce, "'I Have Made a Highway of Biainili': Transportation and Road Networks in the Territories of the Urartian Kingdom," *Altorientalische Forschungen* 48 (2021): 221–32 (especially 224 and Fig. 1).

104 T. A. Sinclair, *Eastern Turkey: An Architectural and Archaeological Survey* (London: Pindar, 1987), 1:178.

Unlike most Muslim chroniclers, who place Alp Arslan in Aleppo before the battle, al-Fāriqī notes that the sultan was in the Iraqi city of Mosul at the time and that he then made it for the town of Arzan in the Diyar Bakr, on the right bank of the Tigris River, before coming to Bitlis, on the southwestern side of Lake Van.[105] Al-Ḥusaynī also places Alp Arslan in Aleppo before his march to Manzikert, but the chronicler remarks that he heard the news of the approach of Romanos's army while in the environs of Khoy and Salmas, a traditional mustering region for nomadic armies on the northern side of Lake Urmia.[106] Most Muslim chroniclers also mention that Alp Arslan dispatched Nizam al-Mulk to Hamadan in Iran, and perhaps to Isfahan as well to set about raising a fresh army to fight the Byzantines.

Matthew of Edessa's account is equally meagre on the geographical and topographical information for the Battle of Manzikert. The *Chronicle* only includes the strategic target of Aleppo and Hierapolis for Romanos's campaign of 1068, and it contains no significant topographical details of the surrounding countryside and plains of the latter like those in the accounts of Attaleiates and Skylitzes, although Matthew emphasizes the "impregnable ramparts" of Hierapolis that posed a formidable hurdle for the imperial army to overcome.[107] Matthew also notes the capture of *kouropalates* Manuel near Sebasteia in 1070, a city mentioned again for Romanos's campaign of the following year (1071).[108] Matthew's account agrees with the majority of Muslim sources that put Alp Arslan at Aleppo when he heard of Romanos's march to Manzikert, adding that the sultan "passed through the confines of Edessa, unharmed, and went in an easterly direction towards the mountain called Lesun,"[109] although the exact routes taken by both divisions of Alp Arslan's army heading towards the Lake Van region are unknown.

105 al-Fāriqī, in Hillenbrand, *Turkish Myth*, 33–34.

106 al-Ḥusaynī, in Hillenbrand, *Turkish Myth*, 53.

107 Matthew of Edessa, *Chronicle*, 128; Smbat Sparapet, *Chronicle*, 34.

108 Matthew of Edessa, *Chronicle*, 129.

109 Matthew of Edessa, *Chronicle*, 133; Smbat Sparapet, *Chronicle*, 37. Nicolle believes that Matthew's Mount Lesun could be identified with the modern Karaca Dağ, a shield volcano located in eastern Turkey between Şanliurfa and Diyarbakir provinces: *Manzikert 1071*, 48.

Chapter 4

THE GEOPOLITICAL AND MILITARY BACKGROUND TO THE BATTLE OF MANZIKERT

The Turkoman and Seljuk Incursions into Eastern Asia Minor, 1017–1067

The late 1030s witnessed a struggle for supremacy in Khorāsān between the Ghaznavids and the Seljuks who, by taking advantage of a growing Ghaznavid weakness that followed the death of Sultan Mahmud of Ghāzni (r. 998–1030), gradually seized territories formerly administered by the Ghaznavids. Merv went over to the Seljuks in 1037, followed by the cities of Herāt and Nīshāpūr less than a year later. On 8 Ramażān 431/23 May 1040, outside Dandānqān near the city of Merv (in modern Turkmenistan), a force of some sixteen thousand Turkmen led by the Seljuk brothers Ṭuğril Beg and Čağrī Beg, defeated a heavily armed Ghaznavid army under Sultan Masʿūd of Ghāzni (r. 1030–1041), whose combat effectiveness had been badly impaired by famine and drought conditions in the region.[1] Not only did the victory abruptly end Ghaznavid dominion there, as Ṭuğril was proclaimed emir of Khorāsān on the battlefield, but it also opened up Khorāsān to the Seljuks. Rayy and Hamadhan fell within the next three years, thus becoming strategic jumping-off points for further expansion westwards into Azerbaijan and Upper Mesopotamia.

Attacks in eastern Asia Minor by bands of the so-called 'Iraqiya Turkmen began many years before the Battle of Dandānqān, with the Armenian sources dating them to either the year 465 AE (1016–1017 AD)[2] or 467 AE (1018–1019 AD),[3] about half a decade before Vaspourakan was officially surrendered to the Byzantine Empire by Senek'erim-Yovanēs. There is a consensus among historians that these were nothing more than raids conducted by unruly Turkmen to acquire pasturelands and loot,[4] which were "side effects" of the Seljukid infiltration of Azerbaijan and Upper Mesopotamia.[5]

1 Alex Mallett, "Dandanakan, Battle of," in *Encyclopaedia of Islam Online*,, http://dx.doi.org/10.1163/1573-3912_ei3_COM_25849, accessed July 25, 2022; Bosworth, *The Ghaznavids*, 228–30; Turan, *Selçuklular Tarihi*, 97–99.

2 Smbat Sparapet, *Chronicle*, 12.

3 Matthew of Edessa, *Chronicle*, 44.

4 These "Iraqiya Turkmen may appear as rebels who have escaped their overlords" control, although, as Beihammer highlights, "it would be misleading to view the developments in Asia Minor independently of the ideological and political framework provided by the Seljuk Empire. Apart from various campaigns and attempts by the sultans to exert direct influence over the Turks in Anatolia, both the sons of Qutlumush and other Turkish warlords were in one way or another imbued with the dynastic and political concepts of the Seljuk Empire." Beihammer, *Byzantium and the Emergence of Muslim-Turkish Anatolia*, 65–68 and 70.

5 Beihammer, *Byzantium and the Emergence of Muslim-Turkish Anatolia*, 62–68; Leveniotes, *Η πολιτική κατάρρευση του Βυζαντίου*, 131–35; Peacock, *Early Seljuk History*, 53–71; Bosworth, *The*

These 'Iraqiya Turkmen were followers of one of Seljuk's sons, Isra'il/Arslan, who had been taken prisoner by the Ghaznavids and imprisoned in India in 1025, where he died (in 1032), after which his followers migrated to Persian Iraq and Azerbaijan, areas that would serve as an ideal location for their flocks. Ibn al-Athīr notes some two thousand Iraqiya migrated towards Mount Balkan (modern Uly Balkan) in the west of modern Turkmenistan, whilst those reported raiding in Rayy were five thousand strong.[6]

By the early 1040s, we can distinguish different Turkmen groups with varying strategic targets in Mesopotamia and Armenia. In 1038, a group of Turkmen around Lake Urmia is said to have set off for the "territory of the Armenians," a region that can safely be identified with Vaspourakan,[7] and some four years later other Turkmen groups launched their attacks in the Diyar Bakr province, after fleeing the Seljuk troops that conquered Rayy under the leadership of Ibrāhim Yināl, a cousin of Ṭuğril Beg, in the summer of 1042.[8] These raids culminated in a temporary seizure of Mosul by Turkmen groups in late February 1044, a development that led to a local Arab and Kurdish alliance against them, proving the tenuous relationship between these Turkmen groups and the Seljuk leadership.[9]

Once the Seljuks established a foothold into Azerbaijan and Arran,[10] Turkish incursions into Anatolia steadily increased. In the spring of 1044, the Turkish army led by Qutlumush, the son of Isra'il/Arslan, led an offensive against Qirwas Ibn a-Muqallad, the Uqaylid emir of Mosul, but was routed in battle on April 24.[11] Upon their retreat to Azerbaijan through Vaspourakan and the north of Lake Van, however, Qutlumush's troops inflicted a humiliating defeat upon Stephen Leichoudes, the *katepano* of Vaspourakan.[12] Although the defences of Vaspourakan were still solid, this was a significant morale victory for the Turkmen raiders, with their incursions into Anatolia increasingly coming under centralized Seljukid control in order to gain income and strengthen Seljuk prestige vis-à-vis the Abbasid caliphate and the nobility of the recently acquired urban

---

Ghaznavids, 245–46; Dimitri A. Korobeinikov, "Raiders and Neighbours: the Turks (1040–1304)," in *Early Seljuk History: The Cambridge History of the Byzantine Empire, c. 500–1492*, ed. Jonathan Shepard (Cambridge: Cambridge University Press, 2008), 698–99; Sevim, *Anadolu'nun Fethi*, 20–24; Turan, *Selçuklular Tarihi*, 88–92.

6 *The Annals of the Saljuq Turks*, trans. Richards, 13–15.

7 *The Annals of the Saljuq Turks*, trans. Richards, 18; Beihammer, *Byzantium and the Emergence of Muslim-Turkish Anatolia*, 64.

8 *The Annals of the Saljuq Turks*, trans. Richards, 20–21.

9 *The Annals of the Saljuq Turks*, trans. Richards, 21–22.

10 Sevim, *Anadolu'nun Fethi*, 29–30.

11 Leveniotes argues that this battle took place much later, probably in early 1057, and that Skylitzes and Matthew are apparently confused in their accounts. Leveniotes, *Η πολιτική κατάρρευση του Βυζαντίου*, 149–50. Sevim, on the other hand, dates the battle in 1045. Sevim, *Anadolu'nun Fethi*, 28.

12 Skylitzes, *Synopsis*, 420; Matthew of Edessa, *Chronicle*, 74.

centres.[13] Hence, in 1047, one of Ṭuğril Beg's nephews, Hasan "the Deaf," the governor of Herat and Sijistan, launched another invasion from Tabriz to Vaspourakan, thus confirming Ṭuğril's aggressive stance against the Byzantine Empire that was taking shape during this period.[14] It was only Hasan's army, but also the Turkmen who were urged on next year by Ibrāhīm Yināl to invade the imperial territories, penetrated as far west as the valleys of Theodosiopolis and reached Chaldea in the northwest and Taron in the south.[15] Yet, the first battle between the imperial forces and the Seljuks that drew the attention of Byzantine sources in more detail was the Battle of Kapetron[16] (modern Hasankale, Erzurum province; September 19, 1048), where the *doux* of Ani Katakalon Kekaumenos, the *katepano* of Vaspourakan Radomir Aaron, and the imperial *magistros* Liparit IV of Kldekari, indecisively clashed with the Turkish forces of Ibrāhim Yināl some 40 km east of Theodosiopolis.

Ibn al-Athīr writes under the year 440 (1048–1049) that a couple of decisive motives behind the incursions into Anatolia were vocalizd by Ibrāhīm Yināl: "My lands are too small to accommodate you [Turkmen] and to provide what you require. The best plan is for you to go and raid the Byzantines and to strive on the path of God and to gain booty. I shall follow in your tracks and aid you in your enterprise."[17] First, there is a clear reference here to the concept of *jihad* that was still circulating almost a century after the Byzantine Empire's wars of reconquest over the Hamdanids of Aleppo.[18] On top of that, we see the idea of "funnelling" the large number of Turkoman warriors to the alluring target of "the lands of the al-Rūm." This overwhelming need to provide large numbers of unruly warriors with lands, booty, and glory, and their leader having to direct them against a richer target is a common *topos* in the sources that narrate the Norman expansion in Italy in the eleventh century. Indeed, we find the sons of Tancred of Hauteville, whose reasons for venturing to Italy are explained by Amatus of Monte Cassino: "these people [the Normans] had increased to such a number that the fields and orchards were not sufficient for producing the necessities of life for so many."[19] Overpopulation also

13 Beihammer, *Byzantium and the Emergence of Muslim-Turkish Anatolia*, 74–75.

14 Leveniotes, *Η πολιτική κατάρρευση του Βυζαντίου*, 153; Beihammer, *Byzantium and the Emergence of Muslim-Turkish Anatolia*, 75.

15 Skylitzes, *Synopsis*, 423–24; Leveniotes, *Η πολιτική κατάρρευση του Βυζαντίου*, 153–54; Sevim, *Anadolu'nun Fethi*, 30.

16 For the campaign leading up to and for the Battle of Kapetron: Leveniotes, *Η πολιτική κατάρρευση του Βυζαντίου*, 92–97.

17 *The Annals of the Saljuq Turks*, trans. Richards, 67.

18 In order to reinforce his fame as a champion of jihad, Sayf al-Dawla ("sword of the Dynasty"), the founder of the Hamdanid Emirate of Aleppo, who ruled between 945 and 967, used al-Mutanabbī's poetry, a famous Abbasid-era Arab poet at the Aleppan court, as the ideal tool for his jihad propaganda. Al-Mutanabbī's poems belong to a broader category of epic poetry that originated in pre-Islamic Arabia, known as qasida (Arabic for "intention"), al- Mutanabbī, "Poem on Hadath," in *Byzance et les Arabes*, 867–959, trans. A. A. Vasiliev ed. and M. Canard (Brussels: Institut de philologie et d'histoire orientales, 1950).

19 Amatus of Montecassino, *The History of the Normans*, trans. Prescott Dunbar (Woodbridge: Boydell, 2004), bk. 1, 1. See also: Geoffrey Malaterra, bk. 1, 5.

emerges as a reason for pushing Norman youth to abandon their homeland in the last speech that Orderic Vitalis puts into the mouth of Robert Guiscard, the Norman Duke of Apulia and Calabria, a few hours before his death in July 1085: "We were born of poor and humble parents and left the barren country of the Cotentin and homes which could not support us to travel to Rome."[20] The concept of Holy War is also present in the narratives of the Norman chroniclers of the period and, although there was a demographic increase in eleventh-century France, the pressure of population alone can hardly explain why the Normans in particular took the road to southern Italy.[21] I believe the same applies to the Turkomans in Azerbaijan, yet the narrative parallels between the two expansions in eastern Anatolia and southern Italy are intriguing.

Another interesting narrative parallel concerns the concept of the conquering warriors claiming fertile land from "effeminate" defenders. Skylitzes reports that when Qutlumush retreated to Azerbaijan through Vaspourakan, and following his victory over Stephen Leichoudes, he reported back to Ṭuğril and "he mentioned Vaspourakan in passing, saying that it was a highly productive region defended by women, a reference to the soldiers against whom he had fought."[22] A common *topos* in the histories of the Norman expansion in France and Italy is the painting of the picture of the "sexual" conquest of Normandy and Apulia, with the conquered lands portrayed by the Norman historians as beautiful women who freely bestowed their fruits to their new captors, while the Normans are depicted as the quintessential warriors—the definition of bravery and manliness dominating over the "effeminate" and "coward" Greeks and Lombards.[23]

According to the accounts of Matthew of Edessa and Smbat Sparapet, Ṭuğril infiltrated the Vaspourakan region and conquered Perkri in the summer of 1054, slaughtering and capturing the local population. Following an eight-day siege, the inhabitants of Artchesh on the northern shore of Lake Van submitted to him, but it is most likely that the city was back under imperial control after the sultan's withdrawal back to Azerbaijan shortly after. Beihammer and Leveniotes point out that, following Aristakēs's narrative, Ṭuğril's most likely target for that campaign was actually Manzikert, apparently aiming to open the main route that linked Azerbaijan to the Arsanias Valley. Yet, the Turkish failure to quickly undermine the city's defences forced the sultan to abandon the siege,

---

20 Orderic Vitalis, *The Ecclesiastical History of England and Normandy*, trans. Tomas Forester, 4 vols. (London: Bohn, 1853–1856), bk. 7, chap.7, 2:369.

21 Graham Loud, *The Age of Robert Guiscard Southern Italy and the Northern Conquest* (London: Routledge, 2000), 84–85; Georges Duby, "Les 'jeunes' dans la société aristocratique dans la France du Nord-Ouest au XIIe siècle," in *Hommes et structures du Moyen âge: Recueil d'articles* (Berlin: de Gruyter, 2019), 213–25. For arguments against this reason see David Bates, *Normandy before 1066* (London: Longman, 1982), 244–48.

22 Skylitzes, *Synopsis*, 420.

23 Geoffrey Malaterra, *The Deeds of Count Roger of Calabria and Sicily and of his Brother Duke Robert Guiscard*, ed. and trans. Kenneth B. Wolf (Ann Arbor: University of Michigan Press, 2005), bk. 1, 1; Amatus of Montecassino, *The History of the Normans*, bk. 1, 21; William of Apulia, bk. 1, 4, bk. 1,8, and bk. 3, 29; Guillaume de Pouille, *La Geste de Robert Guiscard*, ed. Marguerite Mathieu (Palermo: Istituto siciliano di studi bizantini e neoellenici, 1963). For this study I used Graham Loud's online translation. https://ims.leeds.ac.uk/archives/translations/, accessed May 12, 2022.

fearing the arrival of imperial reinforcements, and head to the weaker fortress-town of Artzkē instead, situated on the northern shore of Lake Van and mid-way between Khliat and Artchesh.[24] But I will return to this campaign in my discussion about the strategic importance of Vaspourakan and the fortresses on the northern shore of Lake Van.

So far, we saw that the Turkish campaigns that penetrated eastern Anatolia between the mid-1030s and the mid-1050s did not culminate in any major conquests for the Seljuks, nor did they undermine the imperial defences in Vaspourakan or in any neighbouring regions. Nonetheless, what these campaigns did achieve was, first, to undermine the Byzantine network of Muslim vassal principalities that functioned as a sort of "buffer zone" in Upper Mesopotamia and northern Syria (I will refer, albeit briefly, to these Arab and Kurdish vassal dynasties in the following section) and, second, to establish the Seljuk state into a supra-regional centralizing power that posed a serious challenge to the dominant geopolitical and military position of both Byzantium and the Fatimids of Egypt.[25] Indeed, Ṭuğril's position as a Seljuk overlord and a Sunni champion of *jihad* was significantly enhanced when, on 2 Shawwal 447/ December 25, 1055, just a few days after entering Baghdad, he took up residence in the "royal palace" and received the u titles of *rukn al-din*, *Šâhanšâh*, and *sultan* from the Abbasid caliph al-Qā'im (r. 1031–1075). Two years later, Ṭuğril was received in audience by al-Qā'im, who conferred on him the new title of *malik al-mashrik wa 'l-maghrib* that handed him authority over the lands controlled by the Fatimid caliphs.[26] Following an alarming rebellion by the Turkish general Arslan al-Basasiri that was crushed in 1058, the Seljuks were left as the single dominant power over al-Qā'im 's caliphate.

Ṭuğril's entry into Baghdad inaugurated a new phase in the Turkish expansion into Asia Minor, as the years 1055–1059 saw the Seljuk invasions gradually spread from Vaspourakan and the Armenian highlands to the more southern border regions along the Anti-Taurus Mountains and the eastern parts of Kappadokia.[27] This strategy was shaped to a large extent by the Byzantine political instability and military weakness that followed the civil strife of 1057 and the devastating Battle of Hades (August 20, 1057), as I will explain in more detail in the following section. Besides Frankish mercenaries, members of the remaining Armenian and Georgian aristocracy also took advantage of the situation by forging military coalitions with Turkish warlords.[28] Therefore, it should not come as a surprise that just two months after the Battle of Hades in October 1057, we find the Turks devastating a vast area as far west as Kamakh (modern Kemah, 50 km southwest of Erzincan) and Melitene, while another group headed towards Koloneia in

24 Leveniotes, *Η πολιτική κατάρρευση του Βυζαντίου*, 154–59. See also: Beihammer, *Byzantium and the Emergence of Muslim-Turkish Anatolia*, 80–84; Sevim, *Anadolu'nun Fethi*, 33–34.

25 Beihammer, *Byzantium and the Emergence of Muslim-Turkish Anatolia*, 83–84.

26 D. Sourdel, "al-Ḳādir Bi'llāh," in *Encyclopaedia of Islam Online*, http://dx.doi.org/10.1163/1573-3912_islam_SIM_3761, accessed July 27, 2022.

27 Beihammer, *Byzantium and the Emergence of Muslim-Turkish Anatolia*, 104–5; Leveniotes, *Η πολιτική κατάρρευση του Βυζαντίου*, 103–5; Sevim, *Anadolu'nun Fethi*, 35–36.

28 Leveniotes, *Η πολιτική κατάρρευση του Βυζαντίου*, 99–104.

the north.[29] Apparently, they were answering a call by Liparit IV of Kldekari, the same imperial *magistros* who had been captured by Ibrāhim Yināl in 1048, for Matthew of Edessa explicitly refers to Liparit's son Iwane as "commander of the infidel forces."[30] The entire region between Melitene and Taron was exposed to Turkish raids as a result, with the looting of Sebasteia by the troops of the emirs Samouch, ʻamr-Kāfūr, and Kijaziz following shortly after August 1059.[31]

In 1062, Ṭuġril Beg ordered his emirs Hurāsān-sālār, ČmČm, and Isulv to march against the small Armenian themes (i.e., the "ελάσσονα αρμενι(α)κά θέματα" established between 948 and 952) of Romanopolis[32] (modern Bingöl) and Asmosaton (some 60 km east of the modern city of Elâziğ), and target the key fortresses of T'llxum, Pałin, and Erkni on the slopes of the Anti-Taurus Mountains.[33] The Seljuks leaders and their men engaged in an orgy of looting and plundering, capturing a significant part of the local population, with Matthew of Edessa admitting that the entire area was "unprepared" and "unfortified" but "prosperous and filled with men and animals"[34]—an alluring target indeed. Beihammer believes that the neglect of the fortresses guarding Romanopolis, the Arsanias Valley and the routes leading to Taron, came as a result of an enhanced Byzantine military presence in the Euphrates region that owed a lot to the stabilizing factor of the Marwānid emir Nasr al-Dawlah (1011–1061).[35] Following Nasr's death in 1061, relations between the Marwānids and the Byzantines broke down, thus enabling Seljukid chiefs to gain a foothold in Diyar Bakr in the immediate years after that.

The year 1064 is (in-)famous for the Seljuk conquest of Ani, the old capital of the Bagratid Kingdom of Armenia that became the capital of the theme of Iberia in 1045. Michael Attaleiates claims that the loss of Ani to the Seljuks was due to an alleged Byzantine provocation, seeking to put the blame on the emperor Constantine X and his appointed governor Pagratios Vxkac'i. According to Attaleiates, an imperial force allegedly attacked a Turkish unit that was supposedly marching through the local imperial territories without causing any damage, with Alp Arslan using this provocation as an excuse to attack Ani with considerable forces in retaliation.[36] However, both geographically and strategically, the Seljuk interest in the Caucasus suggests that the new sultan, Alp Arslan (r. 1063–1072), wished to conquer Armenian territories to cover his flanks for a future invasion into Syria, Palestine and, his main target, Egypt,

29 Aristakēs Lastivertc'i, *History*, 136–37; Smbat Sparapet, *Chronicle*, 27.

30 Matthew of Edessa, *Chronicle*, 92–93.

31 Matthew of Edessa, *Chronicle*, 94–95; Sevim, *Anadolu'nun Fethi*, 36; Leveniotes, *Η πολιτική κατάρρευση του Βυζαντίου*, 105.

32 Kühn, Armee, 63; Leveniotes, *Η πολιτική κατάρρευση του Βυζαντίου*, 209–15. Treadgold dates the establishment of Romanopolis to the reign of emperor Romanos I: *Byzantium and its Army*, 35–36.

33 Sevim, *Anadolu'nun Fethi*, 37.

34 Matthew of Edessa, *Chronicle*, 97–98.

35 Beihammer, *Byzantium and the Emergence of Muslim-Turkish Anatolia*, 109–10.

36 Attaleiates, *History*, 147–49.

but also to satisfy the insatiable demand of the Turkomans for booty, pasturelands, and the pursuit of *jihad* into Christian lands.[37] The importance of the area around Ani for Alp Arslan and his Turkoman warriors has also been suggested by Peacock who, following the comments in the account of Aristakēs about the total devastation inflicted upon the conquered cities, coupled with al-Ḥusaynī's account about the use of *naffatun*[38] by the army of Alp Arslan for that purpose, those being a special unit of troops—predominantly Iranian Daylami—that wore fire-proof suits and threw incendiary materials—naphtha grenades and naphtha emitted from a siphon, led the historian to argue that the Seljuk sultan was only interested in the rich pasturelands of the region rather than acquiring any sort of control over the aforementioned urban centres.[39] However, Beihammer downplays the motive for pasturelands and, instead, puts the emphasis on the pursuit of *jihad* and Alp Arslan's wish to advance his claims to supreme leadership of the Muslim world, bearing in mind that the sultan had inherited all the possessions of the dynasty except Kerman, in southern Iran, less than a year before he marched against Ani.[40] Eventually, the city fell to Alp Arslan and was plundered on August 16, 1064, which represents the northernmost advance of Seljuk troops in Christian-Armenian territories.

Within the next couple of years, Seljuk raiding parties began to penetrate the territories of Aleppo and Antioch, even besieging Nisibis for a short period, with a significant clash reported between Gumüştekin Dānishmend Ahmed Gazi, the legendary founder of the Dānishmend dynasty, and the *doux* of Edessa Aruandanos, near a fortress called Oshen.[41] When the *doux* decided to lure the Turks into a trap by feigning a retreat, his troops proved inferior to the task and the retreat broke down into a disorganized pursuit, with the Roman army being unable to execute even this basic tactic, and "the emir fettered Araundanos by an oxen yoke and in that manner led him into captivity." In 1067, Afşin Bey partnered with Gumüştekin to route the Byzantine garrison of Melitene, capturing and pillaging Kaisarea in the aftermath,[42] while Matthew of Edessa reports that "around this period" Afşin Bey "marched forth with a very great number of troops and wintered at the foot of the Black Mountains [modern Nur Dağlari, 90 km north of Antioch]."[43]

---

37 Leveniotes, *Η πολιτική κατάρρευση του Βυζαντίου*, 111.

38 Ahmad Y. al-Hassan and Donald R. Hill, *Islamic Technology: An Illustrated History* (Cambridge: Cambridge University Press, 1992), 106. For more on the use of special troops throwing incendiaries in the Byzantine and Islamic armies of the tenth century: Theotokis, *Byzantine Military Tactics*, 204–5.

39 Peacock, *Early Seljuk History*, 149–50.

40 Beihammer, *Byzantium and the Emergence of Muslim-Turkish Anatolia*, 112–13.

41 Matthew of Edessa, *Chronicle*, 126–27. "Aruandanos" is probably a reference to the *vestarches* and *doux* of Edessa Leon Arvantinos (Λέων Αρβαντηνός Βεστάρχης και δουξ, also recorded/documented as Arvindinos or Ravendinos) who served ca. 1066–1067: Leveniotes, *Η πολιτική κατάρρευση του Βυζαντίου*, 588n3603.

42 Attaleiates, *History*, 169–73.

43 Matthew of Edessa, *Chronicle*, 125.

The Geopolitical and Strategic Significance of Vaspourakan/Vaasparakania

"If these three cities, Khliat and Arzes and Perkri, are in the possession of the Emperor, a Persian [Arab] army cannot come out against Romania, because they are between Romania and Armenia, and serve as a barrier (φραγμός) and as military halts (απλίκτα) for armies."[44] This important statement was written between 948–52 and was included in Emperor Constantine VII's (r. alone, 945–959) monumental work on the administration of the Byzantine Empire, intended for his son and heir to the throne, Romanos (future Romanos I, r. 959–963). It emphasizes the geopolitical importance of Vaspourakan to the eastern frontiers of the empire, something which becomes more apparent to the reader of the *De Administrando Imperio* if they compare the length of the so-called "caucasian chapters" to the rest of the work, but it also highlights the strategic significance of the fortress-towns of Khliat, Artchesh, Perkri, and Manzikert, on the northern shore of Lake Van, as "buffer zones" between Armenia and the Abbasid Caliphate.

The Byzantine Empire's annexation of Vaspourakan took place either in 1019 or, more likely, in 1021/22, when the emperor Basil II was operating with his army in the East.[45] According to Matthew of Edessa, the first clash of the Armenians with the Turkoman bands of raiders in the year 467 AE (1018–1019 AD)[46] greatly unnerved the Arcruni rulers of the principality, and following the "[solemn] alliance and peace [that] was made between the emperor Basil [II] and the Armenian king Senekerim [Senek'erim-Yovanēs] in the year 449 AE [1000–1001 AD],"[47] the way for Vaspourakan's incorporation into the empire was open. The Arcruni ceded their kingdom to the emperor in exchange for territories in the Euphrates, Tephrike, Sebasteia, and in some eastern parts of Kappadokia.[48]

The new imperial principality of Vaspourakan/Vaasparakania combined all the lands of the Arcruni and their vassal princes around Lake Van, including Βαλαλείσων [Bałēš, modern Bitlīs], Αρτζικέ [Artzkē/Arckē], Αρτζέσιον [ArČēš/Artchesh, modern Erciş], and Περκρί [Berkri/Perkri, modern Muradiye], although Perkri and Artchesh submitted a few years later during the reign of Romanos III (r. 1028–1034).[49] The principality of Vaspourakan/Vaasparakania was organized as a *katepanato* more or less in the same period (early 1020s) as their neighbours to the north and west, the *katepanato* of Iberia, which also included the easternmost *kleisoura* (strategically important dis-

44 Constantine Porphyrogenitus, *De Administrando Imperio*, ed. (Greek text) Gyula Moravcsik, trans. R. J. H. Jenkins (Washington, DC: Dumbarton Oaks, 1985), 44:125–28.

45 Werner Seibt, "Die Eingliederung Vaspurakans in das Byzantinische Reich (etwa Anfang 1019 bzw. Anfang 1022)," *Handes Amsorya* 92 (1978): 49–66.

46 Matthew of Edessa, *Chronicle*, 44–45.

47 Matthew of Edessa, *Chronicle*, 41.

48 Skylitzes, *Synopsis*, 336; Matthew of Edessa, *Chronicle*, 45; Aristakes Lastivertc'i, *History*, 18.

49 Leveniotes, *Η πολιτική κατάρρευση του Βυζαντίου*, 140–41n721–24. For the Mokk' (Greek: *Μωξηγάζ*) principality in the southwest coast of Lake Van: Werner Seibt, "Ταξίαρχος Μωξηγάζ: Ein byzantinischer Kommandant in Mokk' um die Mitte des 11. Jahrhunderts?," *Handes Amsorya* 107 (1993): 145–48.

tricts that included important routes of access and, thus, deserved an independent command and greater autonomy than the usual thematic/provincial subdivisions or *tourmai*) of Manzikert between 1001 and 1019/22, before the latter was incorporated into Vaspourakan/Vaasparakania.[50] Based in the fortress-city of Iwan (Wān, modern Van),[51] the *katepano* was placed in charge of the local Armenian princes, who had their own personal following of elite troops that were to contribute to the defence of the region.[52] The reason behind basing the capital of the new *katepanato* to the east of Lake Van was apparently to monitor and react quickly to the threats that loomed in the east.

Other strategically important fortress-cities on the northern shore of Lake Van, like Perkri, Artchesh, and Manzikert, each acquired their own *strategos* as a subordinate to the *katepano* in Van.[53] Manzikert had already emerged as a strategically important fortress-city a century before the famous battle, when, in the year 417 AE (969–970 AD), Bardas Phokas conquered it and demolished its walls, apparently to eliminate it as a strongpoint for the Arabs,[54] although its walls were restored shortly after by the *emir* of Khliat, Abu Shujā' Badh ibn Dustak.[55] Then, during the reign of Romanos III, a certain Alim "the Saracen" surrendered the fortress-city of Perkri to the empire in the hope that he would be honoured with the title of *patrikios*. When his ambitions went unheeded, the disgruntled Alim "made a secret accord with the local Persians then took the fortress in a surprise night attack, killing six thousand Roman soldiers,"[56] before the empire intervened by dispatching the *patrikios* and *katepano* of Vaasparakania, Niketas Pegonites, "with Russians and other Roman forces" to reclaim the city.[57]

Aristakēs provides a different version of events.[58] First, our historian identifies the "Saracen" commander of Perkri as Xtrik, while at the same time mentions that the then *katepano* of Vaasparakania named Kavasilas, serving ca. 1034–1035, occupied the city

---

50 Leveniotes, *Η πολιτική κατάρρευση του Βυζαντίου*, 66–67, 144–47; Viada A. Arutjunova-Fidanjan, "Some Aspects of the Military-Administrative Districts and of Byzantine Administration in Armenia during the 11th Century," *Revue des Études Arméniennes* 20 (1986–1987): 313.

51 Smbat Sparapet notes Artchesh as the capital of the katepano Stephen Leichoudes, who was defeated by Qutlumush's troops in 1044: *Chronicle*, 21.

52 Leveniotes, *Η πολιτική κατάρρευση του Βυζαντίου*, 142.

53 See the table of names of the doux, katepano, archontes, and strategoi of Vaspourakan/Vaasparakania in: Leveniotes, *Η πολιτική κατάρρευση του Βυζαντίου*, 544–49 (Table B.1.)

54 *The Universal History of Step"anos Tar ōnec'i*, trans. Tim Greenwood (Oxford: Oxford University Press, 2017), 235.

55 Greenwood, *The Universal History of Step'anos Tar ōnec'*, 244.

56 Leveniotes considers these "local Persians" to be Turks. Instead, Shliakhtin argues that "contrary to Attaleiates, John Skylitzes...called the Seljuk Turks exclusively 'the Turks' in his *Synopsis Historion*. Skylitzes (who might have used an unidentified source for the Seljuk story) differentiated them from the Persians (the inhabitants of Persia) and the Saracens (the Arabs)." Shliakhtin, "The Projected Identity of the Turks," 49–50. Moreover, Kaldellis notes that "before the arrival of the Seljuks, then, the term [Persians] was reserved for the ancient Persians or for contemporary peoples and states whose rulers were Iranian." Kaldellis, *Ethnography After Antiquity*, 116.

57 Skylitzes, *Synopsis*, 366–67.

58 Aristakes Lastivertc'i, *History*, 39–40.

of Perkri and installed a cavalry corps there. However, the successor of Kavasilas, whom Aristakēs fails to identify, removed the unit to the more remote district of Arcak, which could meet the supply needs of the cavalry. The situation became complicated when Xtrik, who still controlled the citadel of Perkri, invited his "fellow Persians" to help him get rid of the imperial units there. The Byzantine force stationed in the lower city of Perkri had lost its discipline and had become complacent, while those stationed at Arcak were too far to react in time. Eventually, the "Persian" force of Xtrik managed to annihilate some twenty-four thousand imperial soldiers, a greatly exaggerated figure provided by Aristakēs. This is probably why Skylitzes mentions the dispatch of Pegonites that immediately followed the annihilation of the imperial force, while Aristakēs fails to mention the end of Xtrik and his son.

Regardless of the ethnic origin of these "Persians," we should point out the following: the clash at Perkri should be considered as the first recorded one in our sources on the eastern frontier between the imperial troops and the "Persians." This impending threat could explain the subsequent dispatch of the brilliant soldier and general, the *patrikios* Georgios Maniakes, to replace Kavasilas as *katepano* of Vaasparakania ca. 1036–1037.[59] About a decade later, in 1044, the first considerable Turkish penetration into Vaasparakania took place when the *katepano* Stephen Leichoudes was defeated by an enemy expeditionary force led by Qutlumush, a cousin of Ṭuġril Beg, close to Artchesh on the northern shore of Lake Van. The latter were retreating to Azerbaijan through Vaasparakania after having themselves suffered a defeat by the forces of the Uqaylid emir of Mosul, following which they probably looted the regions around the key fortresses of T'llxum, Pałin, and Erkni that were under the jurisdiction of the *doux* of Mesopotamia,[60] before Leichoudes was forced to react. Only three years later, in 1047, Hasan "the Deaf" launched another invasion from Tabriz to Vaasparakania that confirmed Ṭuġril's aggressive stance against the Byzantine Empire, as mentioned earlier.[61] The Turkish raids went on unabated, acquiring an increasingly destructive character when, in 1048, numerous bands of Turkmen penetrated as far as Manzikert, Theodosiopolis, and even Trebizond in the north.[62]

The strategic importance of the fortress-cities along the northern shore of Lake Van became even more pronounced in the campaigning season of 1054, when Ṭuġril Beg personally led an army into Vaasparakania by the northern shore of the lake, apparently wanting to clear a way into the valley of the Arsanias River (modern Murat River, also known as Eastern Euphrates, it originates near Mount Ararat north of Lake Van, and flows westward for 722 km) and into the valley of the Upper Euphrates through the themes of Taron and Keltzine, as well as the ducate of Mesopotamia. Matthew of Edessa

---

59 Alexios G. C. Savvides, *Γεώργιος Μανιάκης, Κατακτήσεις και Υπονόμευση στο Βυζάντιο του 11ου αιώνα* (Athens: Periplous, 2004), Part B (especially 38–40).

60 Matthew of Edessa, *Chronicle*, 74.

61 Leveniotes, *Η πολιτική κατάρρευση του Βυζαντίου*, 153; Beihammer, *Byzantium and the Emergence of Muslim-Turkish Anatolia*, 75.

62 *The Annals of the Saljuq Turks*, trans. Richards, 67; Skylitzes, *Synopsis*, 423–24; Leveniotes, *Η πολιτική κατάρρευση του Βυζαντίου*, 153–54.

and Smbat Sparapet note that Ṭuğril first overcame Perkri, slaughtering and capturing the local population. He then forced the inhabitants of Artchesh to capitulate, following an eight-day siege during which the inhabitants of the town were reduced to fighting the Turks on the city's streets,[63] since "the people of Artze [Artchesh] would not agree to be enclosed by walls."[64] The town, however, returned under imperial control shortly after the sultan's departure. Aristakēs's account differs markedly with regards to Ṭuğril's strategy for this campaign, and it is worth quoting him in detail:

> Skipping over Artchesh and Berkri, they came and camped near the city called Manazkert in the Apahunik' district, seizing all the extensive places in the fields [i.e., the pasturelands]. [The Sultan] dispatched marauding parties across the face of the land: north as far as the stronghold of the Abkhaz and to the mountain called Parxar to the base of the Caucasus; west as far as the forests of Chanet'ia; and south as far as the place called Sim Mountain. And they seized the entire land as [easily as] reapers working a field.[65]

For the Seljuk campaign of 1054 Aristakēs placed the emphasis on the capture of pasturelands, and this is something I would like to consider again later, yet it seems implausible that Ṭuğril would have bypassed two key fortress-cities in his rear in order to attack Manzikert. Therefore, we should consider Matthew's and Smbat's references about the siege and capture of Artchesh and Perkri as more credible, even though Aristakēs was a contemporary of the events. Fortunately, Beihammer and Leveniotes made a valiant attempt to reconstruct the different stages of Ṭuğril's 1054 campaign by including the critical information from Skylitzes's *Synopsis*.[66] At first, the Turkish troops penetrated into Vaasparakania by the northern shore of Lake Van and, indeed, after ignoring Artchesh and Perkri, marched against Manzikert. Although the defenders of Manzikert seem to have been caught unawares and were not prepared for a long-term siege, Ṭuğril, who apparently had a strong siege train but little siege experience, departed with his army after just three days, a decision that earned him the scorn of Aristakēs ("come and marvel at the Sultan's stupidity").

After marching north and west, Ṭuğril soon realized that Theodosiopolis, the old administrative seat of Iberia, was too strong for a prolonged siege and, to make matters worse, he was informed, according to Skylitzes, that a relief army was assembling in Kaisarea. This news probably forced Ṭuğril to march back to Vaasparakania and to the district of Apahunik' (between the Arsanias River and Lake Van) and Manzikert. By this stage of the campaign, however, the garrison of Manzikert was well-prepared and

63 An eight-day siege according to Matthew, but only a six-day siege for Skylitzes. Matthew of Edessa, *Chronicle*, 86; Skylitzes, *Synopsis*, 424.

64 Matthew of Edessa, *Chronicle*, 86; Smbat Sparaper, *Chronicle*, 25.

65 Aristakēs Lastivertc'i, *History*, 93.

66 Skylitzes, *Synopsis*, 424–26 and 432–34; Aristakēs Lastivertc'i, *History*, 93–107; Beihammer, *Byzantium and the Emergence of Muslim-Turkish Anatolia*, 81–83; Leveniotes, *Η πολιτική κατάρρευση του Βυζαντίου*, 156–59.

the city was skilfully and vigorously defended by the *patrikios* and *strategos* Basil Apokapes, whose force apparently also included a small unit of Franko-Normans. It is easy to imagine that Ṭuğril was unable to prolong the siege of Manzikert because of the fear of the imminent arrival of the relief force from Kaisarea, hence his strategic reaction to invest and eventually take the fortress of Artzkē to the south of the Süphan Daği. Despite gathering vast amounts of loot and captives, Ṭuğril "had been unable to accomplish what he had wanted,"[67] to take the fortress-city of Manzikert that would have allowed him to open the way between Azerbaijan and the valley of the Arsanias River.

Following the Battle of Hades in 1057, Turkmen warrior groups were able to penetrate larger sections of the borders, with the focus now shifting to the "central" frontier belt along the Anti-Taurus Mountains, the Euphrates Valley, and the eastern parts of Kappadokia. The years between 1058 and 1062 saw the entire region between Taron, Melitene, Sebasteia, Keltzine, and Theodosiopolis were exposed to repeated Turkmen raids that must have caused considerable damage, a high degree of insecurity for the local population, and, unavoidably, a decay in agricultural production and trade.

By the middle of the 1060s, Turkmen warlords began to infiltrate the defences of northern Syria and the southwestern section of the Byzantine borderland, targeting the territories of Edessa, Aleppo, and Antioch, while Turkmen activities in Upper Mesopotamia and northern Syria during this period for the first time extended to the southern part of the central Anatolian highlands (i.e., Kappadokia). There may not have been any substantial conquests of urban centres in the aforementioned regions on the eve of the Byzantine campaigns led by Emperor Romanos Diogenes, yet the territories that were under the effective control of the empire between Taron, Melitene, Sebasteia, Keltzine, and Theodosiopolis were dwindling fast, while the Seljuk sultanate, especially as a result of its Caucasian campaign of 1064, had taken decisive steps in opening routes from Azerbaijan to Armenia, central Anatolia, Upper Mesopotamia, and northern Syria.[68] Therefore, we should consider the strategic importance of Manzikert and the rest of the fortress-cities along the northern shore of Lake Van within the wider context of Romanos Diogenes's decision to "plug the gap" in the Byzantine defences and, if possible, to reverse the declining geopolitical and military situation in the empire's eastern borders.

But how can we explain the strategic significance of the northern shore of Lake Van over its southern coast for the Turkoman and Seljuk raiders of the period? The reasons are both topographical and geopolitical. To begin with the latter, the fortifications which the Byzantine Empire "inherited" in Vaspourakan/Vaasparakania in the 1020s were concentrated largely on the southern rather than on the northern side of Lake Van; most of them were built between the Deveboynu Peninsula in the east and the Surb Gevorg (St. George) Monastery of Goms in the west, while others were guarding the approaches to Ostan (Wustan, modern Gevaş) and Tatvan.[69] This large concentration of fortresses in

67 Aristakēs Lastivertc'i, *History*, 107.

68 Beihammer, *Byzantium and the Emergence of Muslim-Turkish Anatolia*, 123–24.

69 Robert H. Hewsen and Christoper C. Salvatico, *Armenia: A Historical Atlas* (Chicago: University

Figure 1: Hoşap Castle, southeast of Van. The surviving structure was built in 1643 by Sarı Süleyman Bey, but it was almost certainly erected near the far more ancient Urartian fortress which, in the medieval period, was an integral feature of the kingdom of Vaspourakan, centred around Lake Van. This fortress was rendered irrelevant when the Turkomans arrived from the north of Lake Urmia. Photo: G. Theotokis.

the south is a strong indication that the defences of the region were orientated to deal with threats from the south rather than from the east, thus rendering them irrelevant when the Turkomans arrived from the north of Lake Urmia.

Moreover, no evidence has come to light to indicate that the Byzantines added significantly to the fortifications of the region following its incorporation in 1021/22, something that will have devastating consequences for the defences of Vaspourakan/Vaasparakania against the Turkoman incursions.[70] Therefore, to appreciate the significance of this network of defences, we need to understand the geopolitical threats and networks of alliances in the region in the decades leading up to the Byzantine annexation of Vaspourakan/Vaasparakania.

The most powerful emirate beyond the empire's northeastern marches were the Shaddadids, a Kurdish dynasty that ruled from about 950 until 1200, first in Dvin and Ganja, later in Ani.[71] It was Fażl I (r. 985–1031) who greatly expanded his emirate's ter-

of Chicago Press, 2001), 117 (Map 93).

70 Peacock, *Early Seljuk History*, 135.

71 Clifford Edmund Bosworth, "Shaddādids," in *Encyclopedia of Islam*, 9:169–70; Andrew Peacock, "Shaddādids," in *Encyclopædia Iranica Online*, https://www.iranicaonline.org/articles/shaddadids, accessed June 29, 2022; Peacock, "The Political and Dynastic History of the Iranian World (A.D.

ritory in Arrān, reclaiming territory in Siunik' (to the south of Lake Sevan) from the Pahlavunis, while he was also able to re-establish Shaddadid rule in Dabīl (Dwin/Τίβιον, modern Dvin) in 1022. A clash with the Georgians led by King George I (r. 1014–1027), probably around 1027, ended in a crushing defeat for Fażl, while at the very end of his reign, in 1030, the Caucasus witnessed a dramatic invasion by the Rus' that added to the chaos in the region. With the Byzantine Empire re-asserting its power over the Armenian princes and the Seljuk Turks gaining influence over Arrān, as we saw earlier, it was shortly after their capture of Ganja in 1054 that the Shaddadids, along with other local rulers like the Kurdish Rawwadids[72] of Tabriz and northeastern Azerbaijan, were forced to become vassals of the Seljuk sultan Ṭuğril. Byzantine attempts to reduce Shaddadid castles around Dvin, in 1046–1047, came to a disappointing end for the empire, yet critically Skylitzes mentions a treaty between Abu'l-Aswār Shavur I (r. 1049–1067) and the emperor Constantine IX (r. 1042–1055) that was violated by Shaddadid raids into Byzantine territory around 1055.[73] An imperial counteroffensive on Dvin early next year was significant enough to compel the emir to renew his agreements with the imperial government.

Much more critical for the empire's eastern frontier regions were the Marwānids that ruled between 983 and 1085.[74] Of Kurdish Sunni origin, the Marwānids were based in the Diyar Bakr region of Upper Mesopotamia, centred on the cities of Amid (modern Diyarbakir) and Mayyāfāriqīn, and ruled over Khliat, Bitlīs, Nisibis, Siirt, Cizre, and Hasankayf. Manzikert had been conquered by Bardas Phokas around 969/70, as we saw earlier, who demolished its walls to render it strategically unimportant for the local Muslim chieftains. However, during the rebellious movement of Bardas Skleros between 976 and 979, Abu Shujā' Badh ibn Dustak (known simply as Baḍ), the founder of the Marwānid dynasty, seized Manzikert and rebuilt its walls in 978/79. The city was captured by David of Tao after the death of Baḍ in 990 and, following the death or assassination of David in 1000/1001, all his lands were ceded to the empire.[75]

Baḍ's story is the epitome of an opportunistic warlord of the period; he was most likely born in Sêrt (Siʿird, modern Siirt), southwest of Lake Van, and he was a chieftain of a local Kurdish tribal federation centred around Sêrt and Bitlīs that was vassal to the Hamdanids of Aleppo.[76] After Skleros had asked for his military assistance against the

1000–1217)," in *The Cambridge History of Iran*, ed. J. A. Boyle (Cambridge: Cambridge University Press, 1968), 5:34–35.

72 Andrew Peacock, "Rawwadids," in *Encyclopædia Iranica Online*, http://www.iranicaonline.org/articles/rawwadids, accessed June 29, 2022.

73 Skylitzes, *Synopsis*, 434.

74 Clifford E. Bosworth, *The New Islamic Dynasties: A Chronological and Genealogical Manual* (Edinburgh: Edinburgh University Press, 1996), 91–92; Carole Hillenbrand, "Marwānids," in *Encyclopaedia of Islam*, 6:626; Paul A. Blaum, "A History of the Kurdish Marwānid Dynasty (983–1085), Part I," *Kurdish Studies: An International Journal* 5 (1992): 54–68.

75 Hrats Bartikian, "Ὁ κουράτωρ Μαντζικὲρτ καὶ τῆς ἔσω Ἰβηρίας καὶ ἡ ἀνατολικὴ πολιτικὴ τοῦ Βασιλείου Β'," *Byzantina Symmeikta* 14 (2008): 61–63.

76 H. F. Amedroz, *The Marwānid Dynasty at Mayyafariqin in the Tenth and Eleventh Centuries A.D.*

imperial government of Basil II,[77] Baḍ grasped the opportunity to raid and loot Taron and sack the city of Muş and, following his capture of Manzikert in 978/79, the rest of the key fortress-cities of Khliat, Artchesh, and Perkri soon followed.[78] It was only in 992/3, following a series of Byzantine punitive raids around Lake Van, that Emperor Basil was able to negotiate a lasting peace with the Baḍ's successor.[79] On top of that, when Basil II came to Lake Van in 1000/1001 to accept the lands ceded to the empire after the death of David of Tao, Baḍ's second son Mumahhid al-Dawla Sa'īd (r. 997–1010) came to offer his submission to the emperor and, in return, to receive the rank of *magistros* and the office of *doux* of the East.[80]

The Marwānids's modest territory and an internal crisis in the year 1011 prevented them from undertaking any great projects for the time being, although they managed to seize Edessa in 1026, only for the city to be captured for the Byzantines by the famous George Maniakes, who became the first *strategos* of Edessa in 1031.[81] It would be the long rule of Nasr al-Dawlah Ahmad ibn Marwān (r. 1011–1061) that represented the apogee of Marwānid power in the region. His skilful accommodation with the great powers in the region, the Būyid emir Sultan al-Dawlah, the Fatimid caliph al-Hakim, and the Byzantine emperors, all of whom acknowledged him as the rightful ruler of the Diyar Bakr, brought stability to the region. When the sultan Ṭuğril advanced into the Diyar Bakr (448/1056–1057), he did not aim at abolishing the Marwānid state, so Nasr al-Dawlah recognized his suzerainty and kept his lands, thus confirming his role as a frontier lord fighting the infidels on behalf of the new Muslim superpower in the region.[82]

Another dynasty fighting for a position in the northern Jazira and the Diyar Bakr, to the south and southeast of Lake Van, were the 'Uqaylids.[83] A Shi'a Arab Bedouin tribal group, the main line of which centred on Mosul, they ruled from 990 to 1096. The dynasty emerged in Mosul after the decline of the Hamdanid power in the region, when the first 'Uqaylid Muhammad b. al-Musayyab held the city as a nominal vassal of the Būyid emir Bahā' al-Dawla. It would be Qirwash ibn al-Muqallad (r. 1001–1050) who would struggle to defend his dominions in face of the mounting Turkoman raids in western Persia and Iraq during the 1030s and 1040s, temporarily losing Mosul in February

(Cambridge: Cambridge University Press, 1903), 123.

77 Werner Seibt, *Die Skleroi: eine prosopographisch-sigillographische Studie* (Vienna: Österreichischen Akademie der Wissenschaften, 1976), 46.

78 Rahimi Tekin, *Ahlat tarihi* (Istanbul: Osmanli Araştirmalari Vakfi, 2000), 35; Clifford Edmund Bosworth, "Aḵlāṭ," in *Encyclopædia Iranica Online*, http://www.iranicaonline.org/articles/aklat-or-greek-khliat-khleat-armenian-khlat-a-town-and-medieval-islamic-fortress-in-eastern-anatolia-in-the-former-, accessed June 29, 2022.

79 Catherine Holmes, *Basil II and the Governance of Empire (976–1025)* (Oxford: Oxford University Press, 2005), 310.

80 Thomas Ripper, *Die Marwániden von Diyár Bakr: Eine kurdische Dynastie im islamischen Mittelalter* (Würzburg: Ergon, 2000), 140–41.

81 Savvides, *Γεώργιος Μανιάκης*, 37–38.

82 *The Annals of the Saljuq Turks*, trans. Richards, 111–12.

83 Bosworth, *The New Islamic Dynasties*, 92–93.

Figure 2: The landscape from Muş to Manzikert and then to Van.
Photo: D. Hendrix. Used with permission.

1044, as we saw earlier. As a Shi'a dynasty, they were natural allies of the Fatimids of Egypt and a natural target of the Sunni Seljuks, yet Quraysh ibn Baraka (r. 1052–1061) eventually had to acknowledge the suzerainty of the Seljuk sultan, although his son Muslim ibn Quraysh (r. 1061–1085) later came into conflict with Alp Arslan when he switched his allegiance to the Fatimids, a change of heart that brought Seljuk armies to Mosul, forcing Muslim to flee to Amid and Aleppo. As Beihammer rightly notes, the 'Uqaylids may have seemed too far off from the strategic centres of the region under imperial control, but, due to their position as emerging geopolitical players in the northern Jazira and adjacent regions, the emperors must have been eager to maintain friendly relations with them.[84]

Moving on to the topographical reasons behind the strategic significance of the northern shore of Lake Van for the Turkoman raids, the great diversity of landscapes in the region gives another strong indication as to why the invaders chose the "northern" rather than the "southern" route. Sinclair's archaeological survey highlights exactly this point, "north, west, and east of the lake [Van] there are plains of varying sizes, separated by chains or zones of large rounded hills and mountains. To the south, the land is steeper. A belt of mountains, the eastern end of the Taurus range, lies immediately south of the lake; the Hakkari group in the region,s southeastern corner, is the greatest concentration of high mountains in Turkey and one of the few ranges in the country to resemble Alpine scenery in the steepness and sharpness of the peaks."[85]

Having travelled to the region a few months before writing these lines, I was struck exactly by the same points mentioned in Sinclair's survey. Driving from Muş to

84 Beihammer, *Byzantium and the Emergence of Muslim-Turkish Anatolia*, 59.

85 Sinclair, *Eastern Turkey*, 1:175.

Figure 3: The north coast of Lake Van. Photo: D. Hendrix. Used with permission.

Figure 4: The south coast of Lake Van. Photo: D. Hendrix. Used with permission.

Manzikert and then to Van, the landscape was broken and rugged with plains of varying sizes and small streams feeding into Lake Van, and with the lush grass making the land far more suitable for livestock farming than for agriculture, while the coastal areas were flat and relatively smooth for many kilometres inland. Driving from Van to Gevaş to take the tourist boat to Akdamar Island, I was also awestruck by the change in the landscape, with the high mountains rising sharply and menacingly at just a few hundred metres off the coast and which were still snow-capped in late June.

Although the southern shore of Lake Van is more mountainous, it had a high concentration of monasteries that offered rich and easy pickings for the marauding Turkoman bands, yet, paradoxically, there is scarcely any record of attacks in this region during this period. Peacock makes a very intriguing suggestion that the Turkoman and Seljuk campaigns were linked to the exigencies of the nomadic lifestyle and the acquisition of seasonal pastures for the summer (high cool *yaylas*) and winter (warmer low-lying *kişlaks*) periods.[86] Hence, let me return to Aristakēs's account that explains Ṭuğril's strategy for his 1054 campaign into Vaasparakania:

> Skipping over Artchesh and Berkri, they came and camped near the city called Manazkert in the Apahunik' district, seizing all the extensive places in the fields [i.e., the pasturelands]. [The Sultan] dispatched marauding parties across the face of the land: north as far as the stronghold of the Abkhaz and to the mountain called Parxar to the base of the Caucasus; west as far as the forests of Chanet'ia; and south as far as the place called Sim Mountain. And they seized the entire land as [easily as] reapers working a field.[87]

Considering the great difference between the Seljuk Turks and their established institutions, and the Turkomans and their long-held nomadic lifestyle and tribal customs, and if we bear in mind that the conquest and settlement of Anatolia was undertaken almost entirely by the Turkoman tribesmen with only very limited involvement by the Seljuks's new *ghulām* troops,[88] it becomes easier to appreciate Aristakēs's comments on the strategy for the 1054 campaign. Nonetheless, the seizure of minor castles that controlled the main routes between Azerbaijan and the targeted region certainly was an important aim for the raiders that cannot be disputed.

Hendy makes the pertinent point that "an accurate map of Byzantine-held territory in Anatolia [in the twelfth century] should take account of both the summer and winter seasons, when the situation may have differed quite sensibly," going on to emphasize that the sedentary understanding of a static frontier did not have the same meaning to the nomadic pastoralists.[89] In fact, the seasonal movement of nomadic populations to their summer pastures between May and October, a time the sedentary populations were harvesting their crops and planting for the new season, and back to their *kişlaks* in winter, was indeed well-known to the Byzantines and occasionally received an imperial response. In 1124, when Emperor John II (r. 1118–1143) "unexpectedly attacked them [Turks] in winter...since they were still untrained in agricultural labours...and were always encamped in scatterings on the plain."[90]

---

86 Peacock, *Early Seljuk History*, 145–51. See also: Michael F. Hendy, *Studies in the Byzantine Monetary Economy c. 300–1450* (Cambridge: Cambridge University Press, 1985), 115–17.

87 Aristakēs Lastivertc'i, *History*, 93.

88 Peacock, *Early Seljuk History*, 73, 128. Contra, Beihammer, "perhaps even more than the Iraqi Turkmens, Tughril Beg's forces in the 1040s must have already included large contingents of Iranian, Kurdish, and Arab warriors and slave soldiers. Hence, it is questionable to what extent the early invaders of the Byzantine provinces still had the character of preponderantly Turkmen nomadic fighters." *Byzantium and the Emergence of Muslim-Turkish Anatolia*, 75.

89 Hendy, *Studies*, 117.

90 Ioannes Kinnamos, *Deeds of John and Manuel Comnenus*, trans. Charles M Brand (New York: Columbia University Press, 1976), 17.

To reiterate Peacock's point about the seasonal movements of the Turkomans in Anatolia, if the need to secure summer and winter pastures is taken into consideration for the Turkoman and Seljuk campaigns of the mid-eleventh century, then the invasion patterns and the general directions these took (always under the leadership of leading members of the Seljuk family)[91] make much more sense. Finally, even though there is no contemporary source to confirm the use of the plains to the north of Lake Van as summer *yaylas*, these were indeed used as such during Mongol times, while Upper Mesopotamia made for an ideal *kişlak*.[92]

Byzantine Weakness?

Many reasons have been adduced for the Byzantine failure to stem the tide of invasions in the period after the death of Emperor Basil II in 1025. The most common cited one is the constant civil wars between emperors and pretenders for the imperial throne which largely have been blamed for the political, economic, and military weakness of the empire in the middle decades of the century. Numerous recalls of eastern forces to the west to suppress seditious movements and engage in civil conflicts greatly weakened the imperial defences in the eastern operational theatre. In fact, in a study about the Byzantines' attitude toward warfare, WarrenTreadgold counted seventy-one civil wars from 1039 to 1391, one every five years, about sixty per-cent of the total in just thirty per-cent of the Byzantine period.[93]

A critical rebellion in the midst of both a foreign invasion and domestic unrest was the one launched by the distinguished general George Maniakes. Sometime after 1036, Byzantium decided to put into action Basil II's plan of ten years before to conquer Sicily, as the infighting in the island between different political factions seemed too good an opportunity to miss.[94] As a commander of the land forces for this campaign, Maniakes had units from the eastern themes of Anatolikon, Armeniakon, Thrakesion, and Opsikion under him, while the fact that the Balkans were stripped of a sufficient amount of troops to crush the Bulgarian tax-revolt of 1040 suggests that many units from the west had joined them as well.[95]

Timing is of paramount importance here, as Maniakes landed in Sicily in the spring of 1038, and we saw how the same year a group of Turkmen around Lake Urmia is said to have set off for the "territory of the Armenians" (i.e., Vaspourakan). Although Maniakes mounted a relatively successful war in Sicily against the local Arabs between 1038 and 1040, capturing many towns in eastern Sicily (including Syracuse), where the Christian population was concentrated, he fell out with the admiral Stephanos and the

91 A critical point in: Peacock, *Early Seljuk History*, 148–49.

92 Peacock, *Early Seljuk History*, 146–48 (especially Figure 5.3 and nos. 128–33).

93 Warren Treadgold, "Byzantium, the Reluctant Warrior," in *Noble Ideals and Bloody Realities: Warfare in the Middle Ages*, ed. Niall Christie and Maya Yazigi (Leiden: Brill, 2006), 226.

94 Alex Metcalfe, *The Muslims of Medieval Italy* (Edinburgh: Edinburgh University Press, 2009), 70–87.

95 Kaldellis, *Streams of Gold*, 171.

parakoimomenos John the Orphanotrophos, and was arrested and imprisoned in the capital between 1040 and 1042. Meanwhile, Turkoman incursions into eastern Anatolia steadily increased, as we saw, while the Normans of Apulia succeeded in establishing themselves in the strategic fortress-town of Melfi, in the Apulian–Campanian borders, in 1041; an event with profound long-term implications for the political *status quo* in southern Italy. They even managed to defeat the local *katepano* three times in a single campaign season (at Olivento on March 17, at Ofanto on May 4, and at Montepeloso on September 3).[96]

Between 1025 and 1041 the Byzantine throne had become increasingly vulnerable, but there still had been no outbreak of open civil war. In April 1042, Maniakes was dispatched by the new government of Emperor Michael V (r. 1041–1042) back to Italy. However, once another government (since June 11, 1042) of Emperor Constantine IX Monomachos (r. 1042–1055) sent a replacement for Maniakes, this did not go down well with the fearsome general, who obviously had enough of being mistreated by successive governments in Constantinople, and decided to kill his replacement in October (1042). The following year, the government of Constantine had to face not just a rebellion by one of its top generals, but a surprise attack against the capital by a foreign power as well.

After declaring himself emperor at Dyrrachium, Maniakes advanced to Thessaloniki, spreading terror before him, but his army was challenged at Ostrovo, in March 1043, in a battle where he was mortally wounded in the *mêlée*. With the imperial authority becoming weaker by the month, one of the empire's most capable generals now dead, and other domestic rivals were being suppressed,[97] Emperor Constantine had to face a serious siege of the capital by hundreds of Rus' longships that arrived at the mouth of the Bosphorus in July.[98] Although there is no evidence of possible collusion between Maniakes and the Grand Prince of Kiev Jaroslav I (r. 1019–1054), the immediate threat to the capital by these two enemies inevitably took the attention of the imperial government away from the operational theatres of eastern Anatolia and Italy.

Between 1043 and 1047, the empire would enter a dangerous phase of war on many fronts. In the Balkans, the most significant geopolitical development was the invasion of imperial territory by Patzinak tribes in the winter of 1046–1047, and a settlement of about eight hundred thousand of them between Sofia and Niš.[99] The fact that the empire had to accept the settlement of these Patzinak tribes on imperial soil and employ them as garrison troops and thematic levies reveals its political and military inability to block their passage south of the Danube. In 1052 and 1059, the Byzantines would be forced to mount two large-scale campaigns to "pacify" them.

Related to the Patzinak invasion of 1047 was a seditious movement mounted by Leon Tornikios, a high-ranking officer from Adrianople and a second cousin of Emperor

96 Theotokis, *The Norman Campaigns*, 111–13.

97 Kaldellis, *Streams of Gold*, 185–86.

98 Simon Franklin and Jonathan Shepard, *The Emergence of Rus 750–1200* (Longman: Harlow, 1996), 216–17; Jonathan Shepard, "Why did the Russians attack Byzantium in 1043?," *Byzantinisch-Neugriechischen Jahrbucher* 22 (1978/9): 147–212.

99 Angold, *The Byzantine Empire*, 37–40.

Constantine who, ca. 1045, was a *patrikios*, *vestes*, and *strategos* of Iberia.[100] Following his removal from command, tonsure, and imprisonment in Constantinople in the spring of 1047, for reasons that remain unclear,[101] Tornikios escaped and, having been contacted by certain Macedonian officers, won the support of the tagmata of Macedonia and Thrace and marched against Constantinople in the autumn of 1047. After failing to gain a quick entry into the largely undefended capital, Tornikios's plans fizzled out within a couple of weeks, causing little damage, but again diverting attention and resources away from the eastern theatre where Hasan "the Deaf" launched another invasion from Tabriz to Vaspourakan earlier that summer. For example, the news of Tornikios's seditious movement led to the recall to the capital of the trusted *strategos* from Dvin named Constantine who, along with Tornikios's successor and *doux* of Iberia Katakalon Kekaumenos, had seized the remote Shaddadid fortresses of Hagia Maria (Surb Mari or Surmārī, just south of the Araxes River), Hagios Gregorios (Surb Grigor), and Ampier, in the region between Dvin and Ani. Neverteless, they had to abandon the siege of the key fortress of Helidonion (Cicernak'berd, just north of Dvin) to come to the emperor's rescue.[102]

Most critical of all, however, was the *stasis* of 1057 that culminated in the Battle of Hades, fought on August 20. Following Theodora's nomination of a non-entity as the new emperor Michael VI (r. 1056–1057), in light of his agreed continuation of the policies of Constantine IX, an insurrection by senior members of the *stratiotikon genos* (military aristocracy), led by Isaac Komnenos, son of the general Manuel Erotikos Komnenos, erupted. Isaac made his name as a successful commander-in-chief (*stratopedarches* and *domēstikos ton scholon*) of the eastern armies between 1042 and 1054,[103] when he was replaced by Theodora with one of her eunuch confidants. Other disgruntled officers that took part in the attempted coup against Emperor Michael VI included the *stratēlates* of the East and later *doux* of Antioch, Katakalon Kekaumenos—the same officer (then *doux* of Iberia) who had come to the rescue of Constantine IX in 1047; Constantine Doukas, married to a niece of the patriarch of Constantinople Michael Keroularios; the veteran *strategos* Nikephoros Bryennios (not to be confused with his later namesakes "Elder" and "Younger"); the *magistros*, *strategos*, and future emperor Nikephoros Botaneiates; and the *doux* of Antioch, Romanos Skleros (brother of Maria Skleraina, former mistress of Constantine IX Monomachos)—the "who's who" of the empire's military families, all of whom had recently been dismissed from their posts. Although the socio-political causes and aspects of the 1057 insurrection have been thoroughly examined (and debated) by scholars for over a century, the military aspects of the *stasis* have been (unsurprisingly) rather neglected, until the recent publication of a high-quality research paper by Georgios Leveniotes that explores different aspects of the Battle

100 Leveniotes, *Η πολιτική κατάρρευση του Βυζαντίου*, 537 (Table A.1.13).

101 Kaldellis, *Streams of Gold*, 194.

102 Leveniotes, *Η πολιτική κατάρρευση του Βυζαντίου*, 88.

103 Konstantinos Varzos, *Η Γενεαλογία των Κομνηνών* [The Genealogy of the Komnenoi] (Thessaloniki: Centre for Byzantine Studies, 1984), 41.

of Hades, including the mobilization of forces, the strategy and tactics, the location of the battlefield, the disposition of enemy forces, the battle manoeuvres, and the outcome and aftermath of the battle.[104]

Based on the reports by Psellos and especially Skylitzes, the rebel leaders managed to assemble an impressive array of units under their command.[105] Isaac Komnenos attracted troops from his own family estates in the theme of Paphlagonia and around Kastamone, with smaller units from the neighbouring themes of Anatolikon, Armeniakon, and Boukellarion also joining the ranks of the rebels. Kekaumenos first gathered around him his personal troops, servants, and relatives from Koloneia, while his army was significantly reinforced by Armenian and Roman units fighting in the northeastern border region, such as the tagma of Chaldea, the tagma of Sebasteia, the tagma of Melitene, the tagma of Tephrike, and other Armenian officers and soldiers from Taron, Iberia, and Vaasparakania. On top of that, Skylitzes notes the presence of "two Frankish τάγματα and one Russian" in the army of the rebels, "who were spending the winter in these areas," meaning the themes of Chaldea and Koloneia, and who had been dispatched to Upper Armenia by Michael VI to fight back a Seljuk raiding party the year before (1056), under the command of the Frankish mercenary leader Hervé Frangopoulos.[106] All in all, the rebels had managed to put together a sizeable army of around ten thousand men that came primarily from the eastern regions of the empire, while most of our sources maintain that Michael VI gathered the bulk of his Balkan army to fight the rebels.[107]

The outcome of the Battle of Hades, that was fought a short distance from Nicaea in Bithynia on August 20, 1057, was, undoubtedly, carnage for both sides![108] Matthew of Edessa argued that "all the Roman forces (e.g., the loyalists) were mercilessly slaughtered by the sword," yet the rebel forces sustained heavy casualties as well, although it is futile to try to put any estimates to the paper. What we can say with certainty, however, is that the outcome of the battle encouraged Turkoman tribes to penetrate even deeper into the old imperial Anatolian themata. It was Iwanē, the son of Liparit IV of Kldekari, who jumped at the opportunity to grab territories for himself, even turning against Theodosiopolis, which he tried to seize by trickery and blockade but failed and, inevitably, called the Turks for reinforcements and shared the spoils with them. Naturally, the latter took full advantage of the opportunity to penetrate unhindered into the western-most Armenian regions, raiding and looting a vast area (roughly) between Theodosiopolis/

---

104 Georgios A. Leveniotes, "'Such Carnage in One Place Had Not Occurred before in Byzantium': The Battle of Hades (20 August 1057 CE) and its Repercussions," in *War in Eleventh-Century Byzantium*, ed. Georgios Theotokis and Marek Meško (London: Routledge, 2021), 37–65.

105 Leveniotis, "The Battle of Hades," 46–48.

106 For Herve's career in the Byzantine army: Georgios Theotokis, "Rus, Varangian and Frankish Mercenaries in the Service of the Byzantine Emperors (9th–11th c.): Numbers, Organisation and Battle Tactics in the Operational Theatres of Asia Minor and the Balkans," *Byzantina Symmeikta* 22 (2012): 143–45n88, for the reference on the presence of the Franks in Upper Armenia in 1056.

107 Leveniotis, "The Battle of Hades," 48–49.

108 Leveniotis, "The Battle of Hades," 54.

Erzurum, Keltzinē/Erzincan, Anzitenē, Khorzenē, and Melitene/Malatya, while another group headed towards Koloneia in the north.[109] Aristakēs's writing gives us a taste of the destruction inflicted upon the local populations: "From the very beginning of that year which we recalled above with woe, troops came from Persia resembling ravenous wolves which, upon encountering a flock unguarded by a shepherd, are not content merely with eating their fill, but try to kill all the flock, so the troops from Persia were not satiated by booty alone, but craved our deaths with voracious appetites. It was impossible for anyone they spotted to escape from their hands."[110]

Historians have taken two measures as signalling a narrative of the fiscal and economic decline of the empire in the eleventh century: the "demobilization" of the thematic army of Iberia, and the controlled debasement of the gold coinage. To begin with a brief description of the latter measure, as this is not the main concern of this study, the imperial treasury had been in an increasingly difficult state since the period of Basil II, with his brother and successor Constantine VIII (r. 1025–1028) rescinding Basil's remission of two whole years' taxation, while the next emperor, Romanos III (r. 1028–1035), instituted a laxer fiscal regime that included the abandonment of the *allylengyon* (the practice of forcing the "powerful" landowners to pay any arrears of taxation of their fiscal unit). Michael IV's fiscal policy is also known for sparking the "bulgarian revolt" and the same emperor has been accused of tampering with the currency. Modern numismatists have reluctantly exonerated Michael IV from the latter charge, accusing Constantine IX of implementing a gradual debasement of the gold coinage from twenty-four to eighteen carats—the simplest way to generate more coins with less bullion and to encourage inflation.[111]

Historians have come up with two explanations for the debasement of the coinage in the eleventh century.[112] The traditional one maintains that it was a solution to a budget deficit that burdened the empire's coffers since the period of the so-called "reconquest of the East' in the late tenth century, had spiralled out of control after the death of Basil II especially following the "extravagant" lifestyle and "title handouts" of Constantine IX and his wife Zoe.[113] Then, Cecile Cecile Morrisson has ingeniously suggested that the debasement should be understood as a systemic reaction to the problems of rapid economic growth which the Byzantine Empire supposedly experienced in the early elev-

109 Leveniotes, *Η πολιτική κατάρρευση του Βυζαντίου*, 190–91.

110 Aristakēs Lastivertc'i, *History*, 122–23.

111 Michael Angold, "Belle Époque or Crisis? (1025–1118)," in *The Cambridge History of the Byzantine Empire c. 500–1492*, 590–91.

112 Kaldellis, *Streams of Gold*, 202–3; Angold, "Belle Époque or Crisis?," 590–92; Angeliki E. Laiou and Cecile Morrisson, *The Byzantine Economy* (Cambridge: Cambridge University Press, 2007), 147–55; Hendy, *Studies*, 3–6, 233–36; Angold, *The Byzantine Empire*, 81–98; Cecile Morrisson, "La dévaluation de la monnaie byzantine au XIe siècle: essai d"interprétation," *Travaux et mémoires* 6 (1976): 3–47.

113 On the view of the fiscal policies of Constantine and Zoe by Psellos, Zonaras, Attaleiates, and Kekaumenos: Demetrios Tsougkarakes, "*Η αυτοκρατορία σε κρίση* και η οπτική των σύγχρονων: μια ανάγνωση των μαρτυριών" [The Empire in Crisis and the Perspective of the Contemporaries: A Reading of the Sources], in *Η αυτοκρατορία σε κρίση*, 277–78.

enth century, which, in turn, created a "liquidity crisis" (i.e., insufficient coinage was in circulation). Therefore, Morrisson maintained, the only solution to the inelasticity of the supply of precious metals was to debase the gold coin. However, my instinct is to follow Angold's view in that "even if they [Constantine IX's civil servants] had an inkling that an inelastic money supply was a barrier to economic growth, they were not likely to consider this sufficient justification for debasing a coinage that had remained more or less unchanged since the days of Constantine the Great."[114]

Therefore, if we accept that the most likely explanation for the debasement carried out by Constantine IX was the fiscal difficulties and burgeoning deficit that his government was facing, how were the armies defending the empire affected? It would be better if we let Attaleiates do the talking again.[115] Thus, when the government of Empress Eudokia had to raise an army to stem the Turkoman raids in Upper Mesopotamia and northern Syria in 1067, Attaleiates gives us a sense of the low morale that prevailed among the soldiers: "Again, stinginess made it all ineffectual and inglorious. For those in power did not want to pay a full salary to men who would have to endure a war and the hardships of battle, and they tried to encourage them to risk their lives by giving them only the tiniest recompense. The result was that they failed in every respect...They [soldiers] raised some meaningless shouts and then dispersed to their homes."[116]

During Romanos's first campaign against Alp Arslan in 1068, Attaleiates again paints an equally bleak picture of the general state of the (thematic) troops, probably echoing the complaints and criticisms he would have witnessed in the imperial camp: "It was something to see the famous units (ταγμάτων) and their commanders now composed of just a few men, and these bent over by poverty and lacking in proper weapons and warhorses. For a long time, they had been neglected, since no emperor had gone on an expedition to the east for many years, and they had not received their allotted money for supplies, and little by little they were being defeated and routed by the enemy because they were in a miserable condition and unprepared to meet an attack."[117]

The controversial reform of the "Iberian army" followed the incorporation of Ani to the imperial domains in 1045, a development that upgraded the region of Iberia from a *katepanato* to the *ducate* of Iberia and Greater Armenia.[118] Probably around the turn of the decade, Emperor Constantine IX decided to disband the army of the *ducate* for political, economic, and military reasons. This was a measure that caught the imagination and attracted fierce criticism from contemporary commentators who, compiling their works many years after the events and having the benefit of hindsight, accused Emperor Constantine of *πλεονεξία* (greediness). According to Skylitzes, "He [Constantine] ran so short of money that he invented all kinds of commerce to increase revenue," going on

114 Angold, "Belle Époque or Crisis?," 591.

115 Speros Vryonis, Jr., "The Eleventh Century: was there a Crisis in the Empire? The Decline of Quality and Quantity in the Byzantine Armed Forces," in *Η αυτοκρατορία σε κρίση*, 23–24, 39–43.

116 Attaleiates, *History*, 175.

117 Attaleiates, *History*, 189.

118 Leveniotes, *Η πολιτική κατάρρευση του Βυζαντίου*, 81.

to relate the imperial deficit with the "disbanding of the Iberian army numbering about fifty thousand by the agency of Leo Serblios and then raised heavy taxes instead of soldiers in those regions," only to conclude that "on account of his [Constantine's] prodigality and pretentiousness that the fortunes of the Roman empire began to waste away."[119]

From the perspective of Skylitzes things were straightforward; Constantine was running short of cash for his building project of the Monastery of St. George of Mangana, something we have seen before when sources hostile to the Macedonian dynasty (*Chronicle of Symeon Logothetes*) falsely accused the emperor Basil I (r. 867–886) of not coming to the aid of Syracuse in 878 because the crews of the imperial fleet were occupied with the building of the Nea Ecclesia in the capital.[120] In the emperor's greediness to acquire more money, he dispatched (equally greedy) tax collectors to Iberia and Mesopotamia to create registries in order to impose taxes that had never been paid in those parts; measures that led to socio-political discontent and the eventual collapse of the imperial defences. The imposition of new taxes would attract the same amount of criticism and scorn in any society and in any period; therefore, it is not difficult to believe Skylitzes's suggestion about Constantine's loathed fiscal measures. But is it fair to accuse Constantine IX of *πλεονεξία*?

For more than a century, modern historians have provided different explanations for Constantine's reform of the Iberian army.[121] Yet, we should stick to the core information provided by our sources: the imposition of "new taxes" on some groups in "Iberia," the extraction of taxes instead of soldiers (i.e., military service), and (according to Attaleiates) the termination of the material support provided to the soldiers from "public lands" (*δημοσίων χωρῶν*). What seems to be the case is that, up until that period (end of 1040s), local Armenian soldiers had enjoyed various tax exemptions, or were rewarded for their services with land or with the monetary revenues generated by the taxation of "public lands." Constantine's officials were tasked with compiling new tax registries, abolishing older tax exemptions and methods of remuneration from "public lands," and, critically, introducing new taxes based on these new registries. In the short term, this reform led numerous Armenian soldiers to financial ruin and forced them to defect to the Turkish side in order to survive, and it is possible that the Seljuk leadership was aware of and was encouraged to take advantage of this reform and step up their raids in the region.[122] In the long term, however, we should understand Constantine's new fiscal measures in Iberia as, what is now called, the "fiscalization of the military service,"

119 Skylitzes, *Synopsis*, 444. Compare with: Attaleiates, *History*, 79; Matthew of Edessa, *Chronicle*, 76.

120 Symeonis Magistri et Logothetae, *Chronicon*, ed. Bengt Martin Staffan Wahlgren, Corpus Fontium Historiae Byzantinae, Series Berolinensis 44 (Berlin: de Gruyter, 2006), 264.

121 Martha Gregoriou-Ioannidou, *Οργανωτικά Μέτρα Του Κωνσταντίνου Θ Μονομάχου, Το πρόβλημα του στρατού της Ιβηρίας* [The Organizational Measures of Constantine IX Monomachos, the Problem of the Army of Iberia] (Thessaloniki: Vanias, 1993), 51–74.

122 Gregoriou-Ioannidou, *Οργανωτικά Μέτρα*, 110–16; Leveniotes, *Η πολιτική κατάρρευση του Βυζαντίου*, 84; Werner Seibt, "Stärken und Schwächen der byzantinischen Integrationspolitik gegenüber den neuen armenischen Staatsbürger im 11. Jahrhundert," in Η αυτοκρατορία σε κρίση, 340–41.

which means, in simple terms, that the obligation of a thematic soldier to do military service could now be commuted to cash and the "military lands" (*στρατιωτικές γαῖες*) would be "freed" to pay full taxes to the state.

This was not an unprecedented measure, as it has been suggested that during the reign of Basil II several thematic units in Asia Minor were disbanded because of their poor performance, with the emperor moving towards the "tagmatization of the themata" and the commuting of the military service.[123] The timing of the introduction of these measures by Basil II and Constantine IX was not a coincidence, as these took place during periods of intense military activity when Basil desperately needed the money to finance his Bulgarian wars that lasted up to 1018.[124] Constantine probably decided to disband the local Iberian thematic defence system and especially the "aristocratic" cavalry of the *azatagundk' hayoc'* because, first, he did not trust them very much, but also to make more efficient use of money to hire professional soldiers for the increasingly vulnerable eastern border regions at the end of the 1040s.[125] Yet, a generic complaint coming from the pen of Aristakēs reads:

> But [Monomachos]...was constantly preoccupied with eating and drinking. He elevated filthy people, and as for those taxes which he collected from all lands, which he should have spent on the needs of the cavalry, to enlarge the cavalry forces [fighting] against enemies, [forces] which, by vanquishing the enemies could have kept the land in peace, (as did the venerable Basil during the fifty years of his reign when no enemy dared enter his territory), those accumulated treasures [Monomachos] squandered on whores, and was in no way troubled by the ruin of the land.[126]

Emperor Constantine was no philander or a fool as portrayed by Aristakēs. Rather, he was an astute politician who was fully aware of the strategic danger of the Turkoman penetration in the eastern theatre of operations, and his measure for the Iberian army (ca. 1049–1052) should, therefore, be seen more as a result than as a cause of the aforementioned penetrations, as the latter had been ongoing for several years. The current scholarly consensus is that the emperor was probably just looking for money to improve imperial defences by hiring professional soldiers rather than having to depend on the armies of the local themata, while at the same time, the old military families that constituted the *stratiotikon genos* (military aristocracy) and a serious threat to his authority would be severely weakened by this measure.[127]

---

123 Jean-Claude Cheynet, "Basil II and Asia Minor," in *Byzantium in the Year 1000*, ed. Paul Magdalino (Leiden: Brill, 2003), 82–88.

124 There are three characteristic examples of the fiscalization of the military service from the katepanato of Italy, in Conversano (980), Bari (1017), and Cannes (1034). Jean-Marie Martin, *La Pouille du VIe au XIIe siècle* (Rome: École Française de Rome, 1993), 703–704.

125 Leveniotes, *Η πολιτική κατάρρευση του Βυζαντίου*, 85; Kaldellis, *Streams of Gold*, 211–12.

126 Aristakēs Lastivertc'i, *History*, 108.

127 Kaldellis, *Streams of Gold*, 211–12; Leveniotes, *Η πολιτική κατάρρευση του Βυζαντίου*, 84–86; Jean-Claude Cheynet, "La politique militaire de Basile II à Alexis Comnène," in *The Byzantine*

This leads us to the final question for this chapter, whether the "tagmatization" of many provincial thematic forces, and the reliance of successive governments on mercenary soldiers, are to blame for the decline in the military effectiveness of the imperial forces in the eleventh century. With regards to the quality and military effectiveness of the Byzantine soldier (i.e., the lower echelons of the military hierarchy) in the eleventh century, both Speros Vryonis Jr. and John Haldon made the pertinent point that, although the number of defeats suffered by the imperial armies after the 1060s is noticeably higher than in the previous decades—something that is confirmed by a rough tabulation of the Byzantine victories and defeats for this period published by Vryonis Jr.—Byzantine armies, in fact, also scored significant victories over various enemies in different operational theatres between the reigns of Constantine IX and Romanos IV.[128] Indeed, the disastrous campaign led by Romanos III in Syria in 1030, coupled with the three defeats sustained in a single campaigning season (1041) in Apulia and, in 1048, at Kapetron, gives credit to the point that the Byzantine armies performed no better before the 1050s. Combine that with Attaleiates's account of the poor state of morale and readiness of the provincial armies in 1068, as we saw earlier, which is generally taken as an accurate representation of a recently-developed situation, then the aforementioned statistics seem all the more understandable. Yet, as of the end of the reign of Romanos IV, Attaleiates records five victories (as opposed to eleven defeats) of the Byzantine armies, and some of these were quite insignificant as both Attaleiates and Psellos remarked, but they were victories nonetheless.[129] As Vryonis Jr. notes, Byzantine defeats were more numerous and more *decisive* (i.e., with long-term socio-political implications between adversaries) than their victories.

With the gradual disappearance of the traditional thematic troops came the massively increased reliance by the central government on mercenary forces, both of foreign and domestic origin. The extensive frontier areas subjected to seven decades of expansionist policy had to be manned by seasoned, professional, and well-trained soldiers, a task for which the thematic troops had proved inadequate and ill-prepared. Full-time professional soldiers were more dependent on their employers in terms of pay than the provincial thematic soldiers and, as a consequence, less prone to be involved in politics. They were especially handy to emperors such as Basil II and Constantine IX, who wished to challenge the pre-eminence of the provincial military magnates, particularly the eastern themata.

The Frankish and Rus' mercenaries also proved to be the cheapest and most efficient solution to a period of increased economic troubles and coin debasement. Since their first appearance outside the walls of Constantinople in 1047, the Franks served the emperors in all operational theatres, most frequently by manning castles and strategic

---

Aristocracy and its Military Function, ed. Jean-Claude Cheynet (Aldershot: Ashgate, 2006), 66; John Haldon, "Approaches to an Alternative Military History of the Period ca. 1025–1071," in *Η αυτοκρατορία σε κρίση*, 68–69; Tsougkarakes, "*Η αυτοκρατορία σε κρίση*," 287; Angold, *The Byzantine Empire*, 62–63. See also: Vryonis Jr., Η *παρακμή του μεσαιωνικού ελληνισμού*, 67.

128 Haldon, "Approaches," 47–48; Vryonis, Jr., "The Eleventh Century," 19–21.

129 Vryonis, Jr., "The Eleventh Century," 20–21.

towns in Armenia, Cilicia, and Bulgaria, crucial frontier regions of the empire. They were predominantly cavalry, numbering a few hundred, and since the early 1050s, had their own leaders, the most renowned of them being Hervé Frangopoulos, Robert Crispin, and Roussel of Bailleuil.[130] The long-established view of them being the main cause for numerous rebellions over the centuries, however, has been challenged.[131] In fact, mercenaries hired by the empire were generally a loyal and effective fighting force, and whenever they "misbehaved" against their employers, it was usually in a political context where native Byzantine troops were also equally implicated.

The Byzantines also assigned great importance to the qualities of a commander/leader, as we see particularly in their military manuals like Maurice's *Strategikon* or Leo VI's *Taktika*, because "a ship cannot cross the sea without a helmsman, nor can one defeat an enemy without tactics and strategy."[132] This was not lost on the authors of our sources for this critical period, with both Attaleiates and Skylitzes emphasizing the critical role of the leader in warfare, either by providing leadership models to emulate like George Maniakes or Katakalon Kekaumenos,[133] or particularly when they make comparisons between "military"' emperors like Isaac I and Romanos IV, and others like Constantine IX who failed to give leadership and to confront the enemies of the Romans.[134] Regardless of his "political agenda," Attaleiates does not mince his words: "It was, therefore, during the reign of this emperor [Romanos IV] that the Romans began to stand up to their enemies, recover their more noble outlook, and organize their resistance. For during previous reigns, from the time when Monomachos and his successors exercized power, apart from mere show and an assemblage of sheer numbers, they never engaged in real combat with anyone. It is clear that, generally, what is accomplished is the leader's responsibility in all matters."[135]

It was under the effective leadership of Isaac I that cohesion and discipline were displayed by Roman troops against the Patzinaks in 1059, earning them a remarkable victory, while Attaleiates's account of the campaigns in the period from his accession up to and including the campaign that led to the Battle of Manzikert, proves that Romanos IV was an able and competent commander, taking care to defend his camps and moving his troops in accordance with logistical needs; this is why he was the preferred choice to marry Eudokia and became an emperor in the first place. Nonetheless, to judge from the contemporary historians, defeats at the hands of the Turks in 1044, 1049, 1064, 1067, and 1070 were a result of either incompetent leadership, inadequate tactical preparation, or underestimating the enemy.[136]

---

130 Theotokis, "Rus, Varangian and Frankish Mercenaries," 125–56.

131 Haldon, *Warfare, State and Society*, 92–93.

132 Maurice's *Strategikon*, 64; *The Taktika of Leo VI*, prologue, §9.

133 Skylitzes, *Synopsis*, 365, 375–76 and 402 (for Maniakes); 407, 412–13 and 425 (for Kekaumenos).

134 Haldon, "Approaches," 51–53.

135 Attaleiates, *History*, 219.

136 Haldon, "Approaches," 57n46.

The combination of determined enemies, poor morale, arrears of pay, and sometimes incompetent leadership made for a toxic mix that led to a frequently catastrophic lack of confidence among both the lower (simple soldiers) and the higher (officers) rank of the imperial armies by the later 1060s. However, I believe that it was a combination of two other factors that can better explain the inability of the empire to stem the Turkoman penetrations of Asia Minor in the decades leading up to the Battle of Manzikert; the strategic arrangements for the empire's defence and, what I would call, the empire's "strategic distractions" in other operational theatres.

With the wars of Basil II coming to an end in 1018, the empire found itself much larger than before the so-called "Wars of Reconquest" in the 950s, yet, paradoxically, this development made it much more vulnerable to external threats than before. Basil II's policy of absorbing the Armenian kingdoms of Kars and Vaspourakan resulted in the substantial erosion of the system of client states that worked as a sort of "buffer zone" protecting the main *Rhomaika* themata in the Anatolian interior, a policy that was described as a "poisoned legacy" by Haldon.[137] New defensive districts were created by successive Byzantine governments that were grouped together into a series of larger commands, known as *katepanata* or *ducates*, with the *ducates* of Chaldea, Mesopotamia, and Antioch created in the 970s, expanded to include the *ducates* of Iberia, Vaspourakan, Edessa, and Ani between 1000 and 1045.[138] These were independent of the local thematic administrations and were manned by seasoned professional troops, with individual units of four hundred to five hundred men mentioned frequently in the sources, while regional armies in the eastern frontier numbered variously from a few hundred to a standard force of from four thousand to five thousand infantry and cavalry.[139] Yet, Haldon has made the significant point that this system's strategic response was fragmented and was designed, primarily, to address regional threats rather than to respond to the need to mobilize for larger offensives, thus leaving it to the emperor (or his *domestikos*) to mount a large scale campaign against any such threats like the Seljuk or the Patzinak penetrations. Examples like Basil's campaign against Aleppo in 995, Romanos III's campaign in Syria in 1030, and the Byzantine commanders' inability to communicate and to agree to an effective strategy to counter the Seljuk invasion by the forces of Ibrāhim Yināl that led to the Battle of Kapetron, only serve to reinforce the aforementioned point. This was a system that, essentially, negated the shock-absorbing capacity of the old thematic system that relied on defence-in-depth and on harassment of the enemy.[140]

Another critical factor that upset the new defensive system that was put in place after the "Wars of Reconquest" was the "strategic distractions" of the mid-eleventh century, which included civil wars, rebellions, and foreign raids/invasions. It is a fact that, throughout the history of the Roman/Byzantine Empire, committed warfare on one front was contingent upon a peaceful situation on the others. Add to that the fact that,

137 Haldon, "Approaches," 63.

138 Haldon, *Warfare, State and Society*, 85.

139 Treadgold, *Byzantium and its Army*, 81–85.

140 Theotokis, *Byzantine Military Tactics*, 52–58.

in this new defensive system we just saw, it was left to the emperor to collect an appropriate force and march against any serious external threat, it is easy to understand that any "strategic distraction" would have made the situation much more complicated and dangerous.

In 1043, the government of Constantine IX had to face not just a rebellion by one of its top generals, but a surprise attack against the capital by a foreign power as well. Sources suggest that the government was hastily compelled to recruit levies and mercenaries with which to defend the capital. Combined with the crossing of the Danube by Patzinak tribes in the winter of 1046–1047, Tornikios's rebellion in the spring of 1047 caused little damage in the largely undefended capital, yet his seditious movement diverted attention and resources from the eastern theatre where Hasan "the Deaf" launched another invasion from Tabriz to Vaspourakan earlier that summer. However, as I highlighted earlier, it was the outcome of the Battle of Hades in August 1057 that greatly contributed to the collapse of the eastern borders of the empire in Vaspourakan and Greater Armenia. Not only did it encourage Turkoman tribes to penetrate even deeper into the old imperial Anatolian themata, but the carnage that ensued that day resulted in the loss of many thousands of Armenian and Roman troops from the northeastern border region who had been recruited by Kekaumenos to fight against the royalist forces of Emperor Michael VI. In this decade, the Byzantines were forced to mount two large-scale campaigns to "pacify" the Patzinaks in the Balkans in 1052 and 1059, all the while the Normans of southern Italy inflicted decisive victories over the *katepano* of Italy in 1041—and over the coalition army of Pope Leo IX at Civitate in 1053. This was a battle that firmly established the Normans as an emerging political power in the region, thus opening the way for their three-year siege and conquest of Bari (the capital of the *katepanato* of Longobardia) on April 16, 1071, only a few weeks after Diogenes had departed the capital for his fateful campaign to the East. By the turn of the decade, the empire faced dire "strategic distraction" in all three major operational theatres (Asia Minor, Balkans, Italy), yet it was several factors that always raised Asia Minor to the top of the strategic priorities of the emperors.[141]

141 Neatly summed up by Hélène Ahrweiler, "L'Asie mineure et les invasions arabes," *Revue historique* 227 (1962): 2. For a more detailed analysis of the economic importance of Asia Minor: Haldon, *The Empire That Would Not Die*, 215–48.

Chapter 5

THE PRELUDE TO THE BATTLE OF MANZIKERT

Diogenes's Campaigns of 1068 and 1069

Emperor Romanos's choice of Hierapolis and northern Syria as the main area of operations for the 1068 campaigning season, rather than, for example, the northeastern ("anti-Taurus") region in Asia Minor, should be attributed to the military pressure brought upon the *ducate* of Antioch by the Seljuk raiding parties led by Afşin Bey, Sunduq al-Turki, Ahmetşah, Türkman, Demleçoğlu Mehmet, Duduoğlu, Serhenkoğlu, and others.[1] These particular groups, but especially the one led by Afşin, were launching increasingly destructive raids both in the frontier regions and in the interior of Asia Minor, as we saw in the previous chapter. Afşin's intention may well have been to disrupt the communications between Edessa and Antioch,[2] considering his victory over the *doux* of Edessa in 1066, and his wintering at the foot of the Black Mountains, only 90 kilometres north of Antioch, the following year.

Attaleiates makes a reference to Alp Arslan "detaching a fairly large contingent that he divided into two parts and encamped in upper Asia, placing the one more to the north and the other to the south,"[3] which should be understood as the sultan's strategic response to the news of Diogenes's campaign preparations in the theme of Anatolikon, assigning large detachments in both the northern (Caucasus and Armenia), and the southern (Edessa and Syria) operational areas. A few lines later Attaleiates also notices that "the Persian army to the north...gave the impression that it was withdrawing before the emperor's advance," hinting at the apparent departure of the main Seljuk army from the Caucasus region.[4] Accordingly, we see Romanos's reaction in his advance southeast to the small theme of Lykandos, to the west of Melitene, instead of marching north to Sebasteia and Koloneia, evidently in preparation for an invasion of Syria.[5]

The strategic and tactical inability of the cumbersome imperial forces to block or intercept the smaller and more mobile Turkish raiding parties became apparent when the emperor had to march back north to Sebasteia to deal with the aftermath of an "utterly unexpected" Turkish raid and pillage of Neokaisarea on the Lykos River, around 160 km to the northwest of Koloneia.[6] In fact, Attaleiates calls Romanos's eight-day

---

1 Sevim, *Anadolu'nun Fethi*, 48.

2 Leveniotes, *Η πολιτική κατάρρευση του Βυζαντίου*, 361.

3 Attaleiates, *History*, 191.

4 Attaleiates, *History*, 191–93. Zonaras asserts that Alp Arslan retreated to avoid battle: *Επιτομή Ιστοριών*, 203.

5 Attaleiates, History, 193; Skylitzes, *Continuation*, 83; Beihammer, *Byzantium and the Emergence of Muslim-Turkish Anatolia*, 138.

6 Attaleiates, *History*, 193; Skylitzes, *Continuation*, 87; Zonaras, *Επιτομή Ιστοριών*, 3:204–5.

pursuit of the returning Turkish band across to Tephrike (modern Divriği) "a truly admirable and memorable" feat,[7] which must have set alarm bells ringing in the ears of the emperor and his officers concerning how deep and with what ease enemy forces had penetrated Anatolia while the main imperial army was preparing for an invasion into northern Syria (from the jump-off point in Lykandos, some 350 kilometres to the south of Neokaisarea). Nonetheless, Attaleiates acknowledges Diogenes as a disciple of Ares, the ancient Greek god of war, to laud his zeal for military action, to exert revenge, and to emphasize the emperor's bravery and determination in relieving the Byzantine provinces from the Turkish threat.

On top of that, Attaleiates reveals the incompetence (or, fear? or, perhaps, inexperience?) of an anonymous military commander (*συνταγματάρχης*)[8] who had been detached earlier in the summer, with "a fairly large contingent of the army" that also included the "Franks, men who were warlike and enjoyed bloodshed," to reinforce the garrison in Melitene and protect the flanks of the army from Afşin's incursions.[9] Yet, the commander utterly failed to fend off the enemy attacks, thus pitting the emperor "into a double war rather than a single one"; an ominous start to a critical campaign. This was repeated in late autumn when Afşin and his raiders marched as far west as Amorion, which they sacked, and the failure of the commander stationed in Melitene to relieve the city dealt another military blow that left the emperor "distressed, but unable to take any further action" because of the onset of winter.[10]

But what are we to make of the strategic outcome of Diogenes's campaign in Syria for the summer and autumn of 1068? Hierapolis was of particular strategic importance because its holder was able to control communications between Antioch, Aleppo, and Edessa. It was also a strategic point of contact with the critical river crossings of the Euphrates, just a few kilometres to the east. Therefore, the Byzantine operations to control the wider region of Hierapolis (including the burning of the village of Katma, and driving the Arab garrison from the fortress of Artah) ensured the direct road communication of the eastern part of the *ducate* of Antioch with the *ducate* of Edessa and the city of Samosata. In this sense, the appointment of the *vestes* Pharasmanios Apokapes, an Armenian noble, as the new imperial governor of Hierapolis, should be acknowledged as a strategic success for the emperor. Even more so, considering the scornful comments by Attaleiates, who was a participant but not a combatant in the campaign, about the pathetic state of the imperial army that marched into Syria that autumn.[11]

Moreover, the emperor's decision not to replace the military commander of Melitene has come under criticism by Antonios Vratimos,[12] unfairly in my view, since we do

---

7 Attaleiates, *History*, 195; Skylitzes, *Continuation*, 87.

8 The tactical unit of a syntagma was made up of around two hundred men and their officers. According to the military treatise "On Strategy" (ca. 880 AD), "Two syntagmata make up a pentekosiarchy led by a pentekosiarch." Dennis (ed.), *Three Byzantine Military Treatises*, 49.

9 Attaleiates, *History*, 195–97.

10 Attaleiates, *History*, 223; Skylitzes, *Continuation*, 95.

11 Attaleiates, *History*, 189–91.

12 Vratimos-Chatzopoulos, "Two Expeditions," 184–85.

not know whether the *syntagmatarches*' tactical decisions were based on fear, incompetence, or simply on the strategic reality on the ground (i.e., marching with just a few hundred men, including the heavily armed "warlike Franks," to intercept a highly mobile and experienced raiding party, especially when he was required to relieve Amorion, some 950 kilometres to the west and with no guaranteed logistics, which may have seemed an unrealistic undertaking). What we do see, however, is the strategic initiative passing firmly into the hands of the Seljuks, with Diogenes and the imperial armies now reacting to and struggling to catch up with the Turkish marauders.

The perceived Byzantine weakness in the "central" border region (i.e., Melitene Sebasteia, and Kappadokia) gave rise to a significant increase of hostile activities in the Upper Euphrates and the Anti-Taurus regions for the following campaigning season of 1069. This escalation was precipitated by the revolt of Crispin (Krispinos), another mercenary commander from Norman Italy who, by 1069, had brought significant regions of the Armeniakon theme under his effective control.[13] Diogenes would have realized that the "central" border region between Melitene and Keltzine needed to be firmly secured, yet as we will see, the Byzantine defences were stretched to the limit and eventually broke down on the face of the Turkish onslaught.

Choosing Kaisarea as his main base for the campaign of 1069, Attaleiates reports that Romanos set out for Larissa (modern Mancilik) in the theme of Sebasteia, around 180 km northeast of Kaisarea, to try and chase away a significant force of Turkmen, who were pillaging the countryside around Larissa, all the way to Melitene.[14] Yet, the fact that Diogenes's army proved unable to catch up and deal a decisive blow only served to confirm what we said earlier about the inflexibility of the imperial forces, something that did not escape criticism from our sources.[15] Whether the emperor was fearful or simply cautious of the Turkish tactics, we will never know.

Attaleiates is also highly critical of the emperor's strategic decision to divert his attention to the small theme of Keltzine with only a fraction of his army, having left the rest in the area of Romanopolis under Philaretos Brachamios. This decision to leave the large part of his army behind may have been influenced by his experience against the retreating raiders from Larissa, perhaps taking a chance at going after the enemy with a smaller contingent while the bulk of his forces would reinforce the key fortified towns in the region of Melitene. Instead, Attaleiates strongly advised marching east to Khliat by Lake Van, a move that would have allowed the emperor to try and stem the flow of raiders penetrating eastern Anatolia through the critical entry point of the Arsanias River valley.

Eventually, Diogenes's decision proved to be a strategic blunder that gave the enemies opportunity roam the region and inflict a crushing defeat on Brachamios's troops, before penetrating deep into Kappadokia as far as Ikonion, which they

---

13 Vratimos-Chatzopoulos, "Two Expeditions," 186–208; Theotokis, "Rus, Varangian and Frankish mercenaries," 148–49.

14 Attaleiates, *History*, 229–33; Zonaras, *Επιτομή Ιστοριών*, 3:208–9.

15 Attaleiates, *History*, 233; Skylitzes, *Continuation*, 101.

looted.[16] Attaleiates points out that the absence of Romanos played a critical role in the defeat of the imperial forces under Brachamios, "for the Romans were already obsessed with fear of the Turks, since they had been separated from the emperor and the rest of the host and exposed to the assaults of the enemy...they marched after the emperor, abandoning those places which they had been entrusted to defend."[17] Again, the aforementioned comment confirms what I mentioned in the previous chapter about the poor morale of the Byzantine soldier of the period, coupled with the decisive role of the leader to help suppress the fear of the enemy and to inspire the soldiers to victory. Attaleiates does not mince his words when he writes that "there is nothing that will get the Romans set for war except when they may have to fight against one another."[18]

Attaleiates intervened again in the strategic decision-making process and convinced Romanos to concentrate the army's efforts on the defence of Kappadokia, the defences of which had not yet completely disintegrated.[19] Nonetheless, the emperor's forced march to Heraklea (modern Ereğli in Konya province), in the "Cilician Gates," more than 500 km west of Melitene, could not prevent the escape of the looters of Ikonion into Syrian territory. In fact, had the Turks not been tipped off by Byzantine prisoners of war about the set-up of an ambush at Mopsuestia, Romanos's plan to block their retreat may have worked.[20] However, the Turkish raiders once again managed to elude the Byzantines and escape almost unscathed.

The emperor had failed for a second year to inflict a decisive defeat on a Seljuk army, while the sack of Ikonion was added to that of Amorion the previous year. Furthermore, the "central" border region remained wide open to Turkish raiding parties, with Romanos falling short of winning control over the strategic invasion routes of the Arsanias River valley and the Upper Euphrates region around Melitene. This allowed for an increased number of enemies to roam free between Vaasparakania, Taron, and the Diyar Bakr province south of the Anti-Taurus Mountains. Attaleiates's reported rumour that "other Turks, divided into clans and bands, were raiding through Roman territory, plundering and destroying everything in their path"[21] is no exaggeration and is an indication of the precarious situation on the ground. Therefore, the campaign of 1069 should be considered a strategic failure for Diogenes because it did not reverse the situation prevailing in the late 1060s.

---

16 Attaleiates, *History*, 245–47; Skylitzes, *Continuation*, 101–3; Zonaras, *Επιτομή Ιστοριών*, 3:208–9.

17 Attaleiates, *History*, 245.

18 Attaleiates, *History*, 247.

19 Attaleiates, *History*, 247–51.

20 Attaleiates, *History*, 251; Skylitzes, *Continuation*, 105.

21 Attaleiates, *History*, 251–53.

The Struggle for Supremacy in Syria and Upper Mesopotamia in 1070

The defence of Kappadokia became the main strategic objective of the campaign led by the *protoproedros* and *kouropalates* Manuel Komnenos in 1070, with the emperor seemingly dropping his attempts to seal off the "central' border region of Taron, Vaasparakania, and Melitene. Setting up his headquarters in Kaisarea,[22] Manuel's capabilities in defending central Anatolia and intercepting any invading force was undermined from the beginning when Diogenes ordered him to dispatch a significant part of his forces south to relieve the besieged population of Hierapolis. In fact, since the summer of 1069, it had become apparent to the Byzantine side that the entire region of the emirate of Aleppo faced socio-political instability generally as well as various threats from the armies of Alp Arslan and his emirs. As such, the emperor was desperate not to lose whatever gains he had made in 1068.

In brief, the threats that endangered the stability of Byzantium's "southern" border region came from two directions. Following the easing of the suffocating pressure that had been brought upon the *ducate* of Antioch by the Turkish raids led by Afşin Bey, the duke of Antioch[23] was able to campaign against and occupy the key fortress of Hisn Asfūna, situated in the vicinity of Ma'arrat al-Nu'man, some 80 km southwest of Aleppo, on May 26, 1069.[24] This may well be linked to the Byzantine advance against Artah and Hierapolis the previous year that had aimed to secure a safe zone around Antioch and re-establish communications between Antioch, Edessa, and Samosata. However, the *emir* of Aleppo Mahmūd bin Naşr launched a retaliatory attack against the fortress, besieged it, and annihilated the Byzantine garrison. Eventually, a settlement was agreed that included the destruction of the fortress.[25]

In the late autumn of 1069, the Seljuk *emir* Sunduq al-Turki invaded the region of Aleppo and plundered all the way south to Ma'arrat al-Nu'mān, including the central Syrian towns of Hamah, Hims, and Rafaniyya.[26] Early in the winter of 1069, however, Sunduq withdrew from the region, but not before imposing his suzerainty on the helpless Mahmūd bin Naşr who was anxious to avoid an open conflict. The critical point

22 Attaleiates, *History*, 253–55; Skylitzes, *Continuation*, 107.

23 Based on a comment by Kamāl al-Dīn, historians have assumed that a certain Pext (in Armenian) or al-Naht (in Arabic) was the duke of Antioch in 1069/70 instead of Chatatourios: Leveniotes, *Η πολιτική κατάρρευση του Βυζαντίου*, 602n3678 and 3679.

24 Beihammer, *Byzantium and the Emergence of Muslim-Turkish Anatolia*, 143; Leveniotes, *Η πολιτική κατάρρευση του Βυζαντίου*, 343; Thierry Bianquis, *Damas et la Syrie sous la domination fatimide (359–468/969–1076), essai d'Interprétation de chroniques arabes médiévales* (Damas: Instiiut français de Damas, 1986/89), 2:589.

25 Marco Miotto, *Ο Ανταγωνισμός Βυζαντίου Και Χαλιφάτου Των Φατιμιδών Στην Εγγύς Ανατολή και η δράση των Ιταλικών πόλεων στην περιοχή κατά το 10ο και τον 11ο αιώνα* [The Competition between Byzantium and the Fatimid Caliphate in the Near East and the Activity of Italian Cities in the Region in the Tenth and Eleventh Centuries] (Thessaloniki: Kentro Byzantinon Ereunon, 2008), 240–41.

26 Beihammer, *Byzantium and the Emergence of Muslim-Turkish Anatolia*, 144; Leveniotes, *Η πολιτική κατάρρευση του Βυζαντίου*, 344; Bianquis, *Damas et la Syrie*, 2:590–91.

concerning this development in Aleppo and the wider region is that it demonstrated to the Byzantine side that enemy forces were still able to roam free in northern Syria and Upper Mesopotamia, and ride too close for comfort to Antioch, despite Diogenes's campaign the previous year. This perhaps raises the question whether the conquest of Hierapolis and its neighbouring strongholds did, indeed, secure this part of the empire's defence system. We need to bear in mind that both Afşin Bey and Sunduq al-Turki were leaders who acknowledged the suzerainty of Alp Arslan and acted as a sort of the sultan's "officers" in Upper Mesopotamia and Syria.

It is in this geopolitical context that we should place Alp Arslan's Syrian campaign of 1070/71. But this context can also help us to better understand the emperor ordering Manuel Komnenos to dispatch a significant part of the army under his command in Kaisarea to the "exposed" stronghold of Hierapolis not far from the left bank of the Euphrates. However, what followed Manuel's involvement in the intra-dynastic conflict within the Seljuk ruling house was a devastating invasion of a band of Turkish invaders that reached the westernmost point of the Turkish incursions in Asia Minor, pillaging the city of Chonai in the Upper Meander Valley, "news that greatly depressed us, for it was as though these disasters were being caused by divine anger."[27] In 1070, the Byzantines may well have won over a capable military commander in the face of Chrysoskoulos, but Alp Arslan made an impressive demonstration of prestige and military power that was impossible to ignore.

It is widely believed that Alp Arslan's primary strategic target was not the Byzantine Empire but the Shi'a Fatimid caliphate.[28] Therefore, the Seljuk campaign of 1070/71 laid the foundation for a major invasion of Egypt to follow, with the focus being none other than the city of Aleppo. Sometime around the summer of 1070, most likely in view of the principality's internal weakness and the various external threats, Alp Arslan had some diplomatic contacts with the emir of Aleppo, who around that time ordered that the name of the Fatimid Caliph al-Mustansir cease to be mentioned in the khuṭbah in Aleppo; an act which was tantamount to renouncing any form of allegiance to the Fatimid state.[29] But was the emir prepared to swap the loose allegiance to Cairo for the more demanding one to the Seljuks?

What is certain is that the Seljuk sultan wished to have his flanks covered for the critical campaign in Syria, and for that reason he subdued Artchesh and Manzikert, sometime in the early autumn of 1070. Attaleiates reports that Manzikert was subsequently garrisoned by a strong contingent of Daylamis from northern Iran, while the incessant Turkish raids had transformed the area stretching between Theodosiopolis and Man-

---

27 Attaleiates, *History*, 257; Skylitzes, *Continuation*, 107–9; Zonaras, *Επιτομή Ιστοριών*, 3:210–11.

28 Beihammer, *Byzantium and the Emergence of Muslim-Turkish Anatolia*, 152; Leveniotes, *Η πολιτική κατάρρευση του Βυζαντίου*, 48–50, 287; Nicolle, *Manzikert*, 32; Miotto, *Ο Ανταγωνισμός*, 248–49; Angold, *The Byzantine Empire*, 42–43; Sevim, *Anadolu'nun Fethi*, 51; Turan, *Selçuklular Tarihi*, 169; Bianquis, Damas et la Syrie, 2:593; Claude Cahen, "La campagne de Mantzikert d'après les sources musulmanes," *Byzantion* 9 (1934): 623.

29 Miotto, *Ο Ανταγωνισμός*, 248.

zikert into *άοίκητον* (neglected) *και ἠρημωμένη* (desolate).[30] Following his short but sweeping success in Vaasparakania, Alp Arslan marched towards the Upper Tigris Valley between Amida (Diyār Bakr) and Mayyafariqin, the heartland of the Marwānid emirate, on his way to his first objective of Byzantine-ruled Edessa. The sultan failed to capture the fortress of Amida, forcing instead the Marwānid *emir* of Diyār Bakr, Nasr al-Dawlah Ahmad ibn Marwān, to pay him tribute and provide supplies for his troops.[31] Then, he headed southwest to the border district controlled by the strategic fortress of Sewerakk' (Arab. al-Suwaida, modern Süverek), midway between Diyār Bakr, Samosata, and Edessa, from where he also extracted a thousand dinars in tribute.[32]

Alp Arslan then turned against Edessa, bringing with him a "tremendous army"[33] and his vassals, the Shaddadid *emir* of Dvin and Dabīl, Al-Fadl ibn Shavur (or Fadl II; r. 1067–1073), and the 'Uqaylid *emir* Muslim ibn Quraysh (r. 1061–85).[34] Alp Arslan's army besieged Edessa in mid-winter (1070/71), lasting between thirty and fifty days, but failed to capture it due to the stubborn resistance of the formidable Byzantine garrison and populace (some eighty thousand Armenians, twenty thousand Assyrians, six thousand Greeks, and one thousand Franks),[35] and the competent command of the Bulgarian-born *proedros* and *duke* of Edessa, Basil Alousianos.[36] According to Matthew of Edessa, the Turks unsuccessfully used a hastily built wooden *helepolis* which failed to undermine the city's walls and they resorted to extensive looting of the outskirts of the city, destroying the vineyards and gardens of the Edessans, and using the felled trees to fill in the moat in front of the city walls. After an alleged agreement with the Edessans to pay him 50,000 dinars to end the siege broke down, Alp Arslan decided to withdraw on January 19, 1071, and marched his army across the Euphrates towards the much larger prize of Aleppo.

The sultan arrived outside Aleppo sometime in early February 1071, apparently in an attempt to assert his suzerainty over Mahmūd of Aleppo, insisting that the personal act of submission be performed by Mahmūd in front of his high officials in his camp. Alp Arslan also reminded the emir of his predecessor's title of *malik al-mashrik wa 'l-maghrib*, conferred on Ṭuğril by the Abbasid caliph al-Qā'im on 2 Shawwal 447/ December 25, 1055, which handed him authority over lands controlled by the Fatimid caliphs.[37] If the emir proved to be unwilling to oblige, then he would have to be "persuaded" through the use of military pressure. As a result of the breakdown of negotiations between the two sides, however, the Seljuk army set up camp outside the walls of Aleppo on April 3, 1071, preparing for a full-scale attack on the city. Eventually, just a

30 Attaleiates, *History*, 271.

31 *The Annals of the Saljuq Turks*, trans. Richards, 169; Bar Hebraeus, *Chronography*, 246.

32 Bar Hebraeus, *Chronography*, 246.

33 No doubt, an exaggerated comment: Matthew of Edessa, *Chronicle*, 130.

34 Leveniotes, *Η πολιτική κατάρρευση του Βυζαντίου*, 286.

35 Turan, *Selçuklular Tarihi*, 171.

36 Matthew of Edessa, *Chronicle*, 131–32. For Basil Alousianos: Leveniotes, *Η πολιτική κατάρρευση του Βυζαντίου*, 588 (Table ΙΓ.1.217)

37 *The Annals of the Saljuq Turks*, trans. Richards, 169.

few weeks after the start of the siege (perhaps on May 4, 1071) Mahmūd bin Naṣr and his mother were reduced to submitting to Alp Arslan, with the keys of the city over to him, and the sultan confirming the emir in his position as ruler of Aleppo.[38]

While the sultan was pursuing his strategic goal outside Aleppo, he received an embassy from the emperor Romanos regarding a truce and an exchange of cities. According to Bar Hebraeus, the Byzantines were under the impression that Alp Arslan was aiming towards re-capturing Hierapolis, so the imperial envoy offered to hand over the city to the Seljuks in exchange for Manzikert and Artchesh, and pay an annual tribute to the Seljuk sultan.[39] It is unclear whether the sultan accepted the Byzantine offer because, as we will see, this agreement was overtaken by the events that followed. Leveniotes and Sevim argue that the real purpose of the Byzantine embassy was to mislead the sultan in thinking that the Byzantine strategic plans for the following campaign season included Syria rather than Vaasparakania, which sounds like a very reasonable deception on behalf of Diogenes.[40]

To sum up, Alp Arslan's campaign in Syria in the winter of 1070/71 should be considered a strategic success because it significantly boosted his influence over the regions of northern Syria and the Upper Euphrates by imposing Seljuk suzerainty over the Mirdasid *emir* of Aleppo. It also placed further pressure on the remaining imperial outposts of Edessa and Antioch. The sultan performed a necessary show of force in the regions of his other vassals, the Marwānids and the Uqaylids, and before all this had taken place, he inflicted yet another significant blow to the defences of the Byzantine Empire in Vaasparakania by capturing Manzikert and Artchesh. However, Alp Arslan's attention would be drawn away from Syria by events taking place further north.

Diogenes's Campaign of 1071

To say that Diogenes's ultimate campaign got off to an ominous start would be an understatement, considering that this was an age when superstition and omens played a very real and crucial role in society. It is likely that the emperor would have learned of the fall of Bari, the capital of the *katepanato* of Longobardia, to the Normans and Robert Guiscard on April 16 while he was leading his army through central Anatolia. But even before this disheartening news reached him, several alarming events were recorded by subsequent chroniclers. I mentioned in a previous chapter that later Muslim historians place much more emphasis in their narrative to the meeting and discussion between Romanos and Alp Arslan after the battle, hence there is an interesting episode reported by Sibṭ Ibn al-Jawzī (d. 1256) that the emperor allegedly described to the sultan in detail following his capture:

> The custom is current that a king going out of Constantinople, if he wants to go to war, enters the big church [Hagia Sophia] [and] prays there through

38 *The Annals of the Saljuq Turks*, trans. Richards, 169–70.

39 Bar Hebraeus, *Chronography*, 246.

40 Sevim, *Anadolu'nun Fethi*, 56; Leveniotes, *Η πολιτική κατάρρευση του Βυζαντίου*, 288.

> the intercession of a golden cross, studded with rubies. I entered the church because of this campaign on which I was resolved and I prayed to it. And there was the cross which had fallen from its position in the direction of the Islamic qibla. I was amazed at that and I re-arranged it towards the east. The following day I came to it and there it was inclining towards the qibla. So, I ordered it to be bound in chains. Then I entered on the third day and there it was inclined towards the qibla. I saw [this] as an evil omen and I realised that I would be defeated. Then passion and greed overcame me, I went to the lands of Islam and there befell me what befell me.[41]

Penned by a renowned Muslim preacher and historian during a period of high-religious and propagandistic charge because of the crusades (first half of the thirteenth century), this may well be the author himself talking—Sibṭ Ibn al-Jawzī. Alternatively, the theme of the cross turning towards and, thus, acknowledging the qibla, could have been one current in popular folk stories in Syria, especially in the crusading period.[42] Coupled with the episode of the ritual humiliation of Romanos when the sultan allegedly cut off the emperor's long hair—a symbol of strength and virility in many cultures—and placed his face on the ground, certainly confirms the highly emotional and propagandistic tone of the narrative.

Attaleiates relates other grim omens that foretold the disaster that was about to befall the emperor. First, "while he [Romanos] was crossing the straits of Chalkedon, a dove happened to fly overhead. It was not completely white, but most of its appearance was dark...coming to rest right in his hands."[43] The symbolizm is clear-cut here: a white dove signifies purity and innocence, and, most commonly, protection and peace. On the other hand, it was commonplace for dark doves to appear during times of anxiety or depression, representing transition. Additionally, the emperor crossed the Bosphorus and alighted, not at the customary points of Pylai (modern Yalova) or Nea Kome, but at Helenopolis (Ἑλενόπολις), an ancient Greco-Roman town in Bithynia (identified with the modern village of Hersek, in Yalova Province). Helped by Attaleiates's benefit of hindsight, the town was allegedly nicknamed Eleinopolis, or "miserable city," by soldiers.[44] The collapse of the central pole that supported the imperial tent in the camp was also interpreted as an ominous sign. However, Romanos's decision to sail to Helenopolis instead of Pylai, Greek for "gates," may be related to Emperor Heraclius's decision to land there in 622, at the beginning of his counter-offensive against the Persians.[45] Or was it simply a coincidence?

---

41 Sibṭ *Ibn al-Jawzī*, 72.

42 Hillenbrand, *Manzikert*, 86n148.

43 Attaleiates, *History*, 261.

44 Attaleiates, *History*, 263.

45 "Pylai," in *Oxford Dictionary of Byzantium*, 3:1760; W. M. Ramsay, *The Historical Geography of Asia Minor, Royal Geographical Society, Supplementary Papers IV* (London, 1897; repr. Amsterdam, 1962), 4:201.

On March 13, 1071, the Feast of Orthodoxy, Emperor Romanos crossed the Bosphorus for his last expedition east. Putting a number on the size of the army that the emperor led to Manzikert is futile considering, what I argued in a previous chapter, that the Christian primary sources leave readers frustrated and disappointed in terms of the frequency and accuracy of numbers provided for Romanos's campaigns. Muslim sources, on the other hand, give tantalizingly more details about numbers and estimate of sizes for both armies. Nevertheless, even their accounts vary significantly and their discrepancy in numbers added to the myth that developed around the victory of Alp Arslan at Manzikert.

Nonetheless, some sources do indicate the make up of the imperial army that marched east so, at least, we can have a rough idea about the units that accompanied the emperor. Foreign units that were supposed to serve as a firm foundation for less reliable troops included a detachment of a heavy Frankish cavalry of unknown size, led by the Frank named Krispinos, who was sent to Abydus early in the campaign "on a charge of mutiny";[46] a German mercenary battalion serving as the emperor's personal guard, called the Nemitzoi, whose behaviour and loyalty will also come under scrutiny;[47] a contingent of Franks under another mercenary leader of Frankish origin, Roussel of Bailleuil, again of unknown size but probably no more than one thousand;[48] and detachments of Patzinak and Oghuz Turks, whose exact number is unknown.[49]

Muslim sources like al-Bundārī, al-Athīr, and al-ʿAdīm refer to the presence of "Russian" mercenaries of unknown numbers in the imperial army, which probably indicates battalions from the Varangian Guard.[50] Although the Varangians are not mentioned by any of our Christian sources for the campaign of 1071, a detachment from the guard accompanied the emperor in his 1068 campaign,[51] therefore it is reasonable to assume they would have accompanied Diogenes in 1071 as well. We need to bear in mind, however, that their forces were overstretched in different operational theatres; for example, in the summer of 1070, units from the guard were withdrawn to Italy to assist with the

46 Attaleiates, *History*, 229, 311.

47 Attaleiates, *History*, 269; Skylitzes, *Continuation*, 115; Zonaras, *Επιτομή Ιστοριών*, 3:213.

48 Attaleiates, *History*, 271; Skylitzes, *Continuation*, 115; Zonaras, *Επιτομή Ιστοριών*, 3:213; al-Jawzī, in Hillenbrand, *Turkish Myth*, 38 (gives the exaggerated number of thirty-five thousand Franks); al-Ḥusaynī, in Hillenbrand, *Turkish Myth*, 53; al-Bundārī, in Hillenbrand, *Turkish Myth*, 59; al-Athīr, in Hillenbrand, *Turkish Myth*, 64.

49 Attaleiates, *History*, 271; Skylitzes, *Continuation*, 115; Zonaras, *Επιτομή Ιστοριών*, 3:215; al-Jawzī, in Hillenbrand, *Turkish Myth*, 38 (gives the exaggerated number of fifteen thousand Oghuz); al-Ḥusaynī, in Hillenbrand, *Turkish Myth*, 53; al-Bundārī, in Hillenbrand, *Turkish Myth*, 59; al-Athīr, in Hillenbrand, *Turkish Myth*, 64; al-ʿAdīm, in Hillenbrand, *Turkish Myth*, 74.

50 al-Bundārī, in Hillenbrand, *Turkish Myth*, 59; al-Athīr, in Hillenbrand, *Turkish Myth*, 64; al-ʿAdīm, in Hillenbrand, *Turkish Myth*, 74. Nicolle notes that "Arab and Persian chroniclers added Rus' probably meaning the Varangians, Khazars, Alans, Kipchaks, Persians, Georgians and Abkhazians from the Caucasus." *Manzikert*, 40. Hillenbrand, however, argues that "Rus was the medieval Arabic term used for the peoples living in the areas now known as Russia, Ukraine and Belarus." *Manzikert*, 80n47.

51 Skylitzes, *Continuation*, 85.

defence of Bari, and sources refer to a naval engagement that summer, when a Byzantine fleet comprising Varangian marines defeated Robert Guiscard's inexperienced fleet off the coasts of Brindisi.[52]

Matthew of Edessa writes about the

> very great and formidable number of troops [that] was gathered together from the entire country of the Goths, from all the Bulgars, from all the distant islands, from Kappadokia and from all of Bithynia, from Cilicia and Antioch, from Trabizond, and [last but not least] from Armenia...moreover, Diogenes had mercenaries from the infidels of Khuzestan, as numerous as the sands of the sea.[53]

Using the term "Goths,"[54] Matthew probably means either the Franko-Normans or, possibly, the German Nemitzoi; the troops "from all the Bulgars" probably refers to units from the *katepanato*[55] of Bulgaria; the "distant islands" may also apply to Crete, Cephalonia, and/or Cyprus; finally, the infidels of "Khuzestan" (a historic region in southwestern Persia) should be a reference to the Oghuz troops under Diogenes's command.

None of the Byzantine sources offer any details about the battle array of the imperial army before the battle itself, except Bryennios,[56] who writes that, on the eve of the main battle on Friday, August 26, the left wing of the imperial army comprised mainly of the western tagmata under Nikephoros Bryennios ("the Elder"), and that the right wing included the units from Kappadokia, with the rest of Asia Minor under Theodore Alyates. Based on the aforementioned note, we may assume that Romanos had called for further reinforcements from his western tagmatic units that would, in theory, have raised the number to around twelve thousand men,[57] although their numbers in the field would have been much lower (perhaps even as low as one thousand men each according to Haldon)[58]. According to Bryennios, the emperor also placed Andronikos Doukas in command of palatine units of "the tagma of the *Hetairiae* and that of the nobles" (τάς τε των έταίρων τάξεις έχων και τας των άρχόντων). I have found no reference of the presence of the *Scholai* or the *Stratelatai* in the campaign of 1071,[59] which is noteworthy

52 Georgios Theotokis, "The Varangians of Byzantium," in *Το Βυζάντιο και η Ρωσία του Κιέβου (882–1240)* [Byzantium and Kievan Russia (882–1240)], ed. Georgios Kardaras (Athens: Institute of Historical Research, 2021), 64–65n40.

53 Matthew of Edessa, *Chronicle*, 132.

54 The only other reference to the "Goths" in Matthew's work is for the armies under Tornikios in 1047. *Chronicle*, 75.

55 Reference to the katepano Bulgarias, Andronikos Philokales (ca. 1064/75) found in Kühn, *Die byzantinische Armee*, 227–30.

56 Bryennios, *Ύλη Ιστορίας*, 85–89; Attaleiates, *History*, 287–93; Skylitzes, *Continuation*, 121–23; Zonaras, *Επιτομή*, 215–17.

57 Treadgold, *Byzantium and its Army*, 116.

58 Haldon, *Byzantine Wars*, 171.

59 Haldon, *Byzantine Wars*, 172.

especially considering that the latter had been badly mauled outside Hierapolis in 1068 as we saw earlier. Nonetheless, I am at a loss as to who these *ἄρχοντες* were.

For the indigenous eastern thematic *tagmata* of Asia Minor, we have scattered references to detachments from Armeniakon, Kappadokia, and Anatolikon (units from Pisidia and Lykaonia are mentioned), as well as the Armenian heavy infantry from Theodosiopolis. It is very likely that the neighbouring ducates of Antioch, Chaldea, Sebasteia, Charsianon, and Koloneia also contributed troops to the imperial army.[60] Trying to assess the exact numbers of these units seems hopeless, with the margin of error being high, but in theory the numbers for the small themata like Lykaonia were eight hundred men, while larger ones like Armeniakon could have provided a contingent of five thousand men.[61] Therefore, the grand total of troops that Romanos brought with him in the early summer of 1071 can be estimated at between thirty to forty thousand men,[62] with the majority of them being indigenous rather than foreign mercenaries.

Diogenes's strategic target seems to have been the fortresses along the coast of Lake Van, namely Khliat and Manzikert. But how straightforward is it to map the imperial army's march from the Bithynian coast to the outskirts of Manzikert? The communications network and road system in Anatolia is, of course, subject to geophysical and climatic constraints and, as it was the case in all pre-industrial societies, maintaining it was very irregular since the seventh century or earlier.[63] These long-distance roads, as well as many regional and secondary roads, were, as in most parts of the Roman Empire, generally designed for vehicular traffic drawn by pack animals.[64] Consequently, there evolved a series of major military routes, the upkeep of which was in the state's interest. Along these routes a string of fortified posts and military bases also developed.[65]

---

60 Bryennios, *Ὕλη Ιστορίας*, 85–89; Attaleiates, *History*, 287–93; Skylitzes, *Continuation*, 121–23; Zonaras, *Επιτομή*, 215–17.

61 Treadgold, *Byzantium and its Army*, 81–86.

62 John Haldon, Vince Gaffney, Georgios Theodoropoulos, and Phil Murgatroyd, "Marching across Anatolia: Medieval Logistics and Modeling the Mantzikert Campaign," *Dumbarton Oaks Papers* 65/66 (2011–2012): 212–13; Leveniotes, *Η πολιτική κατάρρευση του Βυζαντίου*, 167; Nicolle, *Manzikert*, 40; Haldon, *Byzantine Wars*, 171; Haldon, "La logistique de Mantzikert," in *Guerre et société au Moyen âge: Byzance-Occident (VIIIe–XIIIe siècle)*, ed. Dominique Barthélemy and Jean-Claude Cheynet (Paris: Association des amis du Centre d'histoire et civilisation de Byzance, 2010), 11–26 (especially 14–15) See also: Turan, *Selçuklular Tarihi*, 177; Sevim, *Anadolu'nun Fethi*, 54–55.

63 Johannes Koder, "Historical Geography," in *The Archaeology of Byzantine Anatolia: From the End of Late Antiquity until the Coming of the Turks*, ed. Philipp Niewohner (New York: Oxford University Press, 2017), 9–27.

64 Klaus Belke, "Roads and Routes in Northwestern and Adjoining Parts of Central Asia Minor: From the Romans to Byzantium, with Some Remarks on their Fate during the Ottoman Period up to the 17th Century," *Gephyra* 20 (2020): 79–98 at 79–83; Belke, "Transport and Communication," in *The Archaeology of Byzantine Anatolia*, 28–38. For pack animals and wheeled vehicles: Richard W. Bulliet, *The Camel and the Wheel* (New York: Columbia University Press, 1990).

65 Johannes Koder, "Regional Networks in Asia Minor during the Middle Byzantine Period: An Approach," in *Trade and Markets in Byzantium*, ed. Cécile Morrisson (Washington, DC: Dumbarton Oaks, 2012), 147–75. The classical study is: Ramsay, *The Historical Geography of Asia Minor*,

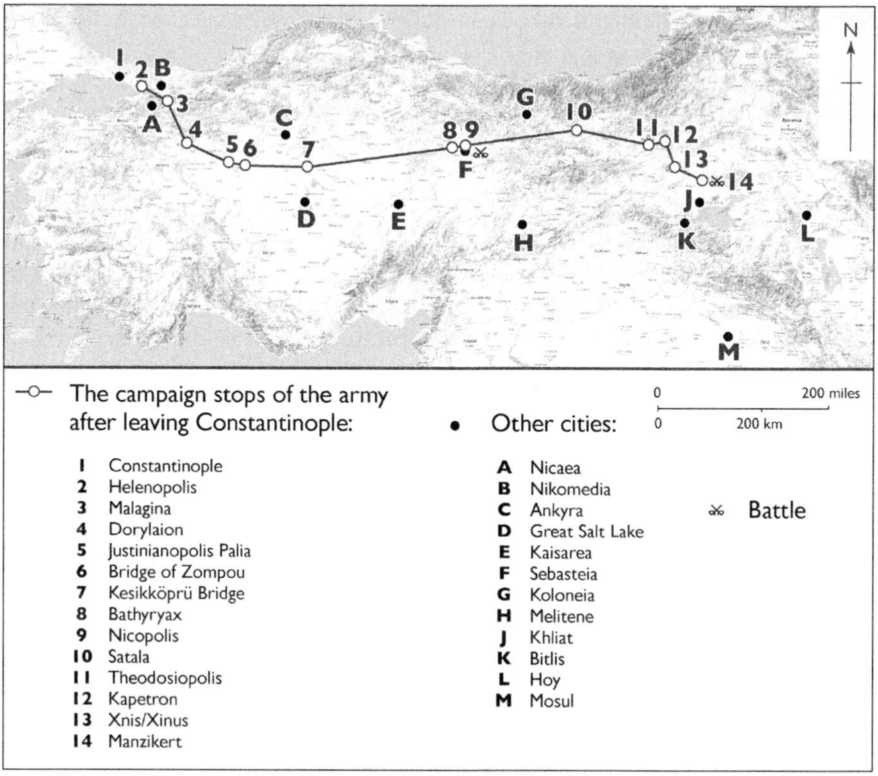

Map 1: The Manzikert campaign of 1071. Map by Cath D'Alton. Map data © Google 2023.

The main military route east from the Bithynian coast went via Nicomedia and Nicaea to the major imperial military bases (ἄπληκτα) at Malagina, and from there to Dorylaion. Malagina (just north of the modern village of Paşalar, in the district of Pamukova; 40 km east of Nicaea and 50 km southeast of Nikomedia) was the first *aplēkton* on the road to the eastern frontier where the strategoi of Thrakesion and Opsikion would gather to join the emperor.[66] First mentioned in the sources for Empress Eirene's campaign against the Arabs in 786, Malagina was a base of considerable strategic importance in the lower valley of the Sangarios River, and a place where imperial stables supplied the imperial armies with mounts for their campaigns.[67] The main military road from Malagina south

4:51–88. See also the series of volumes published as part of the project Tabula Imperii Byzantini by the Austrian Academy.

66 Constantine Porphyrogenitus, *Three Treatises*, 80–81.

67 "Malagina," in *Oxford Dictionary of Byzantium*, 2:1274; Clive Foss, "Byzantine Malagina and the Lower Sangarius," *Anatolian Studies* 40 (1990), 161–83. The remains of Malagina/Metabole have been identified as a Komnenian fortress rebuilt in 1145 after a Turkish raid. Nikos D. Kontogiannis, *Byzantine Fortifications: Protecting the Roman Empire in the East* (Barnsley: Pen & Sword, 2022), 145–46.

to Dorylaion passed through a village named Gaita (unidentified in a modern location), and then to another village identified as Leukai (modern Lefke/Osmaneli, Bilecik province) on the Sangarios where, according to Ramsey, the Bridge of Pithekas was the crossing point at the junction between the Sangarios and the Göksu rivers. This was in the neighbourhood where the castle identified as *Armenokastro* ("castle of the Armenians") would have stood.[68]

Dorylaion (Şarhöyük, near Eskişehir, around 140 km south of Malagina/Paşalar) was the next *aplēkton* for the Byzantine armies. It was one of the "new" economic and military centres and, like Amorion (Hisarköy/Afyon, Turkey) that became the seat of the strategos of the Anatolikon theme, it sprung up in the late seventh century following the Arab expansion. A major military post that controlled a strategic road junction from Constantinople to the interior of Asia Minor, Dorylaion was frequently mentioned after 741, when it became the base of the revolt of Artabasdos. According to Ibn Khordadbeh, Dorylaion was noted for its plains, where imperial pack animals were raised.[69]

After leaving Dorylaion, Diogenes's army would have marched southeast towards the old and very important fortress of Justinianopolis Palia (modern Sivrihisar, Eskişehir province), perhaps the chief city of the Roman province of Galatia Secunda/Salutaris (later incorporated into the theme of Anatolikon), and an important road junction in central Anatolia.[70] The next obstacle to overcome was the crossing of the Sangarios (Sakarya) River that would have taken place at the well-known Bridge of Zompou, one of the main passages used by the Byzantine army to the east, as mentioned by Skylitzes and Attaleiates. The distance between Justinianopolis Palia and the Sangarios is about 40 km as the crow flies and, more than a century ago, Ramsay attempted to identify this particular bridge, dismissing the even earlier theories that it was situated at the headwaters of the Sangarios or at its estuary, and placing it, instead, somewhere to the east of the modern village of Kavuncu, where the Sangarios/Sakarya is joined by the waters of Özdenk Creek.[71]

The main military road then passed to the south of Ankyra, skirting the Great Salt Lake/Tuz Gölü on its eastern side, to join another road from Mokissos (modern Helvadere, Aksaray) to Kaisarea. Before that, however, an army had to negotiate the Halys/Kizlirmak River, with the most likely crossing point being at the Kesikköprü Bridge ("broken bridge") in modern Kirşehir Province, where the surviving bridge was built by the Seljuk Sultan of Rum Kaykaus II (r. 1246–1262) in 1248.[72]

---

68 Refik Arikan, "An Unknown Roman Bridge on Sangarius and Ancient Road System Around It," *Global Journal on Humanities and Social Sciences* 4 (2016): 270–78; Ramsay, *The Historical Geography of Asia Minor*, 4:200–202.

69 "Dorylaion," in *Oxford Dictionary of Byzantium*, 1:655; Ramsay, *The Historical Geography of Asia Minor*, 4:212–13.

70 Ramsay, *The Historical Geography of Asia Minor*, 4:214.

71 Ramsay, *The Historical Geography of Asia Minor*, 4:214–18.

72 Fügen İlter, *Osmanlilara kadar Türk Köprüleri* (Ankara: Karayollari Genel Müdürlüğü, 1978), 167–69.

Figure 5: The most likely crossing point of the Halys/Kızılırmak River was at the Kesikköprü Bridge ("Broken Bridge"), where the surviving bridge was built by the Seljuk Sultan of Rum Kaykaus II (r. 1246–1262) in 1248. © Semih Ekinci. Image available under CC BY-SA 3.0 licence.

At this point, Diogenes's army would have taken a southeastern detour towards Kaisarea, where Attaleiates notes that the emperor denied permission for the bulk of his troops for to enter the city, for obvious reasons, as we saw in a previous chapter. Attaleiates's narrative gives the impression that the emperor decided to move his camp away from Kaisarea, just a few kilometres to the north at a fertile place called Krya Pege ("Cold Spring"), where his troops were able to camp and re-supply themselves for several days. Yet, Attaleiates's narrative is misleading!

It has proven very difficult to pinpoint where this Krya Pege would have been, but there have been some suggestions that it could be identified with Bathys Ryax or Bathyryax (Βαθυρύαξ, "Deep Stream"; modern Kalinirmak pass west of Sivas),[73] around 180 km northeast of Kaisarea and merely 30 km northwest of Sebasteia. This was a location of great strategic importance, as indicated by the fact that it served as a fortified assembly point (*aplēkton*) for the *Armeniaka themata* for the Byzantine expeditions to Greater Armenia,[74] and the place where the emperor Heraclius would also have camped in his campaign of 626.[75] Therefore, Attaleiates may give the impression in his *History* that the emperor avoided pitching his camp on the outskirts of Kaisarea, choosing, instead,

73 Ramsay, *The Historical Geography of Asia Minor*, 4:76; Richard J. A. Talbert (ed.), *Barrington Atlas of the Greek and Roman World* (Princeton: Princeton University Press, 2000), 1:64, 2:987; Thomas S. Brown, Anthony Bryer, and David Winfield, "Cities of Heraclius," *Byzantine and Modern Greek Studies* 4 (1978): 15–38. See also: T. B. Mitford, R. Talbert, T. Elliott, and S. Gillies, "Bathys Rhyax/ Krya Pege?: A Pleiades Place Resource," *Pleiades: A Gazetteer of Past Places* (2012), https://pleiades.stoa.org/places/628944, accessed September 11, 2022.

74 Constantine Porphyrogenitus, *Three Treatises*, 80–81.

75 Brown, Bryer, and Winfield, "Cities of Heraclius," 22–30.

to march to Krya Pege as if this place was a few dozen kilometres away. In reality, and as we have seen, Krya Pege can be identified with a location some 180 km northeast of Kaisarea, a distance that would have been covered in no less than nine or ten days under normal circumstances. Hence, Attaleiates probably identified the places where the imperial army would have encamped for more than a few days to recuperate, reorganize, and stock up on food, and more importantly, fresh water supplies, as it would certainly have been very challenging to find sources of fresh water for an army of thirty to forty thousand men and their pack animals in Anatolia in the early months of the summer. This is exactly what Attaleiates emphasizes in his narrative, "this place [Krya Pege] was lacking in no necessity whatever: the water was crystal clear, drinkable, and very cold; wood was abundant as the land was forested with trees with thick foliage; ... and, so to say; it was a city in the countryside or a rustic city because it was known for the variety and mixture of its resources."[76]

It was at this time when we read about the incident with the German Nemitzoi that resulted in the imperial army's demotion to rear-guard, from their previous place of honour as personal guards to the emperor.[77] Immediately following his aforementioned description of the lush countryside around Krya Pege, Attaleiates noted that "the emperor's soul was stung" when he witnessed the countryside being "looted" by his troops and, indeed, one can imagine why Diogenes forbade his army from camping close or inside the city of Kaisarea a few weeks before. The incident described by Attaleiates involved the emperor leading his troops against (a group of?) Nemitzoi, after some of the latter, "full of insolence, anger, and barbarian mindlessness, recklessly rose up to avenge their own men." What would have happened here is obviously a commandeering of provisions that got out of hand, and consequently, some of the looters would have been punished for their behaviour. Yet, the image of a Byzantine emperor mounting his horse and lining up his army against these men would have been enough to impress them to submission, but it would also have had a negative impact on the army's morale. It may have been around this time that Hervé Frangopoulos, who had probably been campaigning in Amaseia, joined the emperor's army.[78]

While encamped in Sebasteia, Matthew of Edessa reports another worrying incident that would have had an impact on the morale of the imperial army. There, a substantial Armenian population was accused by the Greek-speaking inhabitants of siding with the Seljuks. Matthew of Edessa claims that, despite being courteously received by two local Armenian leaders, Romanos snubbed both of them, as well as the ex-king Gagik, and then ransacked part of the Armenian quarter.[79] About the road system beyond Sebasteia, Attaleiates speaks of two *ἀτραποί* (shortcuts, pathways), one which led southeast to Melitene and the Euphrates, and the other which the emperor took that led some 120 km northeast to Nicopolis (modern Koyulhisar, in Sivas), in the theme of Koloneia; a city

76 Attaleiates, *History*, 267.

77 Attaleiates, *History*, 269; Skylitzes, *Continuation*, 115.

78 Nicolle, *Manzikert*, 42.

79 Matthew of Edessa, *Chronicle*, 132–33; Nicolle, *Manzikert*, 42.

situated in a well-watered plain lying at the base of a thickly-wooded mountain and at a critical road junction where the military roads radiated to cities such as Comana Pontica, Polemonium, Neokaisarea, Sebasteia, etc.[80]

After Nicopolis, the army would have proceeded east towards Theodosiopolis via Satala (modern Sadak, in Kelkit district of Gümüşhane province, some 200 km southeast of Nicopolis),[81] a site known since the times of emperor Trajan as a legionary camp of the Legio XV Apollinaris, and that was re-fortified by Emperor Justinian ca. 530.[82] The distance from Satala to Theodosiopolis, the final stop before entering "enemy territory," would have been around 150 to 170 km. That was a critical juncture in the entire campaign, because our sources emphasize that the incessant fighting and the devastation inflicted in the area between Theodosiopolis and Manzikert forced Romanos to order his troops to carry provisions for two months after leaving the relative safety of Theodosiopolis.[83]

We also saw in a previous chapter that Bryennios provides different background concerning the fateful decision to launch a major strike towards Lake Van, reporting that in the military council that took place sometime during the march between Kaisarea and Sebasteia/Krya Pege, both the respected Georgian *magistros* Joseph Trachaniotes and Nikephoros Bryennios "the Elder" sought to convince Romanos not to take to the field against the Turks, but rather to wait for them in Theodosiopolis. Unsurprisingly, the more bellicose officers in the council prevailed and the march towards Khliat and Manzikert was confirmed, much to Bryennios's bewilderment and consternation. In fact, Romanos and his officers are reported to have been so thoroughly taken with the idea of the "reconquest" of former Roman lands that, according to al-Ṭurṭūshī, whose work contains the earliest extant narrative of the battle but also stories and information from eyewitnesses in circles close to the Seljuk leadership, "They [Byzantines] had divided up the countries of the Muslims—Syria, Egypt, Iraq, Khurasan, and Diyar Bakr—and they did not doubt that the wheel [of fortune] had turned for them and that the stars of good fortune were in their favour."[84] There is little doubt, however, that this is simply propaganda to portray Diogenes as an arrogant and an insolent ruler.

A similar scene was repeated ten years later, when another newly crowned emperor was thrust into the same type of emergency and prevailed upon by his bellicose military council to face an enemy he thought he and his army were prepared for, although they clearly were not; this "usurper" was none other than Alexios I Komnenos (r.

80 Attaleiates, *History*, 268; Belke, "Roads and Routes," 85; Belke, "Transport and Communication," 31 (Fig. 2.1.); Klaus Belke, *Galatien, und Lykaonien*, Tabula Imperii Byzantini 4 / Österreichische Akademie der Wissenschaften. Philosophisch-Historische Klasse: Denkschriften 172 (Vienna: Österreichische Akademie der Wissenschaften, 1984), 104–5.

81 M. Roaf et al., "Satala," *Pleiades* (2018), https://pleiades.stoa.org/places/874684, accessed September 12, 2022.

82 "Stala," *Topos Text*, https://topostext.org/place/396400USat, accessed April 22, 2023.

83 The area is described as "ἀοίκητον" and "ἠρημωμένη": Attaleiates, *History*, 271; Skylitzes, *Continuation*, 115; Zonaras, *Επιτομή Ιστοριών*, 3:212–13.

84 al-Ṭurṭūshī, in Hillenbrand, *Turkish Myth*, 27.

1081–1118), whose army was defeated by the Normans at Dyrrachium on October 18, 1081. Alexios was fortunate to get a second chance against the same enemy and on the same battlefield, some twenty-seven years later, when experience taught him to avoid a pitched battle with the Norman heavy cavalry at all costs. Instead, he chose to blockade the enemy army within its camp at Dyrrachium, an ingenious strategy that bore fruit.[85] Diogenes did not get a second chance; what the emperor desperately needed was accurate information about Alp Arslan's actions and intentions, but he did not get that either.

When it comes to the supply of intelligence to the emperor in this critical juncture of the campaign, with the imperial army stationed at Theodosiopolis and in need of accurate information about the Turkish intentions before marching against the main target of Khliat and Manzikert, the role of Leon Dabatenos has to be studied more closely. Leon Dabatenos (Armenian: Lewon Dawatanos) came from a well-known aristocratic family of Armenian origin that found itself on the rise around the middle of the eleventh century, but was already in decline in the twelfth century.[86] The sources identify a couple of members of the Dabatenos family as holding high-ranking positions in the second half of the eleventh century. A certain Dabatenos was *katepano* or *doux* of Edessa and died during the siege of Amida in 1062.[87] We saw earlier in this chapter that the *doux* of Edessa responsible for the defence of the city from the Seljuks during the month-long siege in the winter of 1070/71 was Basil Alousianos, himself of Bulgarian origin, and the grandson of the *tsar* Ivan Vladislav of Bulgaria (r. 1015–1018), who defended Edessa very effectively against Alp Arslan's army. Yet, Alousianos disappears from this office shortly after the Turkish retreat from the region of Edessa, to be replaced by the *vestarches* and *doux/katepano* Leon Dabatenos who held the office ca. 1071–1077.[88] This change of leadership at such a decisive moment is suspicious, to say the least, but I was unable to identify the reason for Alousianos's dismissal.

Moreover, Bryennios is the only source to account for an official report that reached Diogenes by Leon Dabatenos while the emperor was probably still in Theodosiopolis with the bulk of his army, according to which "the sultan, it read, learning about the campaign [of Romanos to Khliat and Manzikert] and afraid of the power [of Romanos's army], left Persia for Babylon [Baghdad]." But what is more important about the outcome of the campaign is what immediately follows, "the emperor, putting his trust on these words, divided his army in two: he kept one in place [Manzikert], and he dispatched the other to Khliat."[89] This report should be combined with another intriguing comment by Sibṭ Ibn al-Jawzī who writes that "the envoy of Byzantium returned [to the

85 Theotokis, *Norman Campaigns*, 209–14.

86 "Dabatenos, House," in *Encyclopaedic Prosopographical Lexicon of Byzantine History and Civilization: Baanes–Eznik of Kolb.*, ed. Alexios G. Savvides and Benjamin Hendrickx (Turnhout: Brepols, 2008), 2:284; "Dabatenos," in *Oxford Dictionary of Byzantium*, 1:577.

87 Smbat Sparapet, *Chronicle*, 28; Leveniotes, *Η πολιτική κατάρρευση του Βυζαντίου*, 587 (Table ΙΓ.1.211.).

88 Leveniotes, *Η πολιτική κατάρρευση του Βυζαντίου*, 587 (Table ΙΓ.1.218.). Contra: Kühn, *Die byzantinische Armee*, 200.

89 Bryennios, *Ὕλη Ιστορίας*, 83.

imperial camp] rejoicing to his master [Romanos] and that strengthened the resolve of the king of Byzantium to pursue him [Alp Arslan] and to fight him."[90] Do both Bryennios and al-Jawzī basically describe the same person compiling an intelligence report for the emperor that was delivered either in letter (Bryennios) or in person (al-Jawzī)? Was Leon Dabatenos the head of the embassy dispatched by the emperor to Alp Arslan while the latter was in the vicinity of Aleppo in early spring of 1071? There is some food for thought here, but the sources raise more questions than provide answers. Nonetheless, it is very reasonable for the emperor to have demanded intelligence from the *doux* of Edessa about the size, units, and morale of Alp Arslan's army; the critical point here is how much Romanos could trust Leon Dabatenos.

When it comes to the matter of allegiance, technically, Dabatenos's was with the Byzantine emperor, but an extract from Matthew of Edessa's *Chronicle* may suggest otherwise: "The doux [of Edessa], who resided in the city [of Edessa], provided him [Alp Arslan] with horses, mules, and victuals. Taking these, the sultan passed through the confines of Edessa, unharmed, and went in an easterly direction towards the mountain called Lesun."[91] Although he is not mentioned by name, what Matthew suggests here is that the *doux* of Edessa may have acknowledged the suzerainty of the Seljuk sultan, thus having to provide him with the necessary supplies for his march north from Syria to Iran. But what about the report dispatched to Diogenes on the strength and size of the Seljuk army? Did Dabatenos deliberately attempt to mislead the emperor with false intelligence? We will never know with certainty, but the evidence is damning, especially if we consider that he remained in office throughout Michael VII's reign until ca. 1077.

Most Muslim chroniclers place Alp Arslan in Aleppo when he learned of the Byzantine campaign headed towards Lake Van. One exception is al-Athīr who writes that he was already in Khoy, in northwestern Iran, while another is al-Fāriqī who notes that the sultan was in the Iraqi city of Mosul at the time and that he then made it for the town of Arzan in the Diyar Bakr, on the right bank of the Tigris River, before coming to Bitlis, on the southwestern side of Lake Van.[92] Moreover, the only chronicler who reports the day of departure of the sultan from Syria is al-Qalānisī, with the information that Alp Arslan left Aleppo on 23 Rajab 463 AH (April 26, 1071).[93]

Shortly after his departure, Alp Arslan and his army crossed the Euphrates "on horseback without boats."[94] But with the spring thaw in full swing, the melting snows raised the water level in the Euphrates and the outcome was disastrous, as "the riding animals and camels perished, his [Alp Arslan] crossing was like [that of] of the fugitive and he did not pay attention to those people and riding animals that had perished."[95] Alp Arslan knew that his army had been badly mauled by his forced march from Syria

90 Sibṭ Ibn al-Jawzī, in Hillenbrand, *Turkish Myth*, 67.

91 Matthew of Edessa, *Chronicle*, 133.

92 al-Athīr, in Hillenbrand, *Turkish Myth*, 64; al-Fāriqī, in Hillenbrand, *Turkish Myth*, 33–34.

93 al-Qalānisī, in Hillenbrand, *Turkish Myth*, 30.

94 al-Ḥusaynī, in Hillenbrand, *Turkish Myth*, 53.

95 Sibṭ Ibn al-Jawzī, in Hillenbrand, *Turkish Myth*, 67; Matthew of Edessa, *Chronicle*, 133.

to western Azerbaijan, and that is why he "sent the bulk of his exhausted troops home while he led his tough and loyal *askar* of four thousand across the mountains of Kurdistan to Khoy in northwestern Iran."[96] Most Muslim chroniclers also mention that Alp Arslan dispatched Nizam al-Mulk to Hamadhan in Iran and perhaps to Isfahan as well, to set about raising a fresh army to fight the Byzantines.[97] Yet, the exact route that the Seljuk army would have followed during its retreat from Syria remains a mystery. We can only speculate that some of the stops would have included Edessa where, according to Matthew's account (as we saw earlier), the army would have been re-supplied, then northeast to Diyar Bakr and east, perhaps through Bitlis and the southern shore of Lake Van, as the sultan would have been unaware of how far the imperial army would have been from the towns on the northern side of the lake and his flanks would have been covered by the garrisons in Khliat and Artchesh, and then directly east to Khoy in the northwest coast of Lake Urmia.[98] How long would Alp Arslan's preparations have lasted remains another mystery, yet al-Bundārī and al-ʿAdīm recount that the main Seljuk army was on its way to Khliat, though still far away from there, when the first clash between imperial forward units and the Seljuk garrison of Khliat occurred on 4 Dhu'l-Qa'da 463 (August 2/3, 1071).[99]

Estimates of the size of Alp Arslan's army vary considerably to between twelve thousand and twenty thousand strong,[100] although it is only al-Jawzī and Sibṭ Ibn al-Jawzī who report the size of the sultan's *askar* numbering four thousand *ghulams*, while the rest were primarily of Kurdish origin assembled in the regions of the Diyar Bakr and Bitlis.[101] Rashid al-Din claimed that it numbered fifteen thousand cavalry, naming the tribal leaders as Artuq, Saltuq, Mengücük, Dānishmand, Çavli, and Çavuldur.[102] On top

---

96 Sibṭ Ibn al-Jawzī, in Hillenbrand, *Turkish Myth*, 68; al-Jawzī, in Hillenbrand, *Turkish Myth*, 37; al-Ḥusaynī, in Hillenbrand, *Turkish Myth*, 53 (does not mention the disaster in the crossing of the Euphrates); al-Bundārī, in Hillenbrand, *Turkish Myth*, 58.

97 al-Jawzī, in Hillenbrand, *Turkish Myth*, 37.

98 Most of the Muslim sources agree that Alp Arslan's main muster point before marching to confront Diogenes was the city of Khoy: al-Jawzī, in Hillenbrand, *Turkish Myth*, 37; al-Ḥusaynī, in Hillenbrand, *Turkish Myth*, 53 (he adds the city of Sal(a)mas, some 40 km southwest of Khoy, as an additional muster point); al-Athīr, in Hillenbrand, *Turkish Myth*, 64; Mīrkhwānd, in Hillenbrand, *Turkish Myth*, 98–99 (although he adds that the ten thousand troops from Tabriz were led to Vaasparakania by the sultan himself, which is misleading). The only exception is: al-Ṭurṭūshī, in Hillenbrand, *Turkish Myth*, 28 (notes the city of Isfahan as the muster point for the upcoming campaign).

99 August 3, 1071 was a Wednesday. al-Bundārī, in Hillenbrand, *Turkish Myth*, 59–60; al-ʿAdīm, in Hillenbrand, *Turkish Myth*, 75.

100 al-Qalānisī, in Hillenbrand, *Turkish Myth*, 30; al-Fāriqī, in Hillenbrand, *Turkish Myth*, 34 ("only small numbers"); Nishapuri, in Hillenbrand, *Turkish Myth*, 36; al-Jawzī, in Hillenbrand, *Turkish Myth*, 38; al-Bundārī, in Hillenbrand, *Turkish Myth*, 58, 60; al-Athīr, in Hillenbrand, *Turkish Myth*, 64; Sibṭ Ibn al-Jawzī, in Hillenbrand, *Turkish Myth*, 68; al-ʿAdīm, in Hillenbrand, *Turkish Myth*, 74; Rashīd al-Dīn Faḍlullāh Hamadānī, in Hillenbrand, *Turkish Myth*, 90; Mīrkhwānd, in Hillenbrand, *Turkish Myth*, 96.

101 al-Jawzī, in Hillenbrand, *Turkish Myth*, 37; Sibṭ Ibn al-Jawzī, in Hillenbrand, *Turkish Myth*, 69.

102 Rashid al-Din, in Hillenbrand, *Turkish Myth*, 90.

of that, other Turkoman leaders that would have flocked to reinforce the sultan may have included Savtekin, Afşin, Gevher-Âiyn, Tarang/Tarankoğlu, Sunduk al-Turki, and Aytekin.[103] This could easily have taken the total number of troops under Alp Arslan's command to thirty thousand men, including up to fifteen thousand elite cavalry.[104] As we saw in the previous chapter, this discrepancy in numbers between the two opposing armies aims to add to the myth that developed around the victory of Alp Arslan at Manzikert—the myth of David (i.e., Alp Arslan) defeating Goliath (i.e., Romanos Diogenes).

103 Leveniotes, *Η πολιτική κατάρρευση του Βυζαντίου*, 171; Turan, *Selçuklular Tarihi*, 179.

104 Nicolle, *Manzikert*, 49; Leveniotes, *Η πολιτική κατάρρευση του Βυζαντίου*, 172; Sevim, *Anadolu'nun Fethi*, 58.

Chapter 6

THE BATTLE OF MANZIKERT

THE IMPERIAL ARMY under Diogenes was based in Theodosiopolis probably until the high summer of 1071, before advancing into "unfriendly territory" and having to rely on their two-month supplies that the emperor had ordered his troops to carry for the forthcoming march to Lake Van. Yet, it would seem inconceivable that Romanos was unaware that Alp Arslan was assembling an army. As we saw in the previous chapter, we must assume that, in June or July, the Byzantine leader was receiving some very bad or even misleading intelligence reports that would have a disastrous effect on his strategic decisions for the following stage of the campaign.

Strategic Miscalculations

Sometime in late July or early August, Diogenes's overconfidence led him to make a critical strategic error. After leaving Theodosiopolis to march southeast to Manzikert via Xnis/Xinus (modern Hinis, Erzurum province),[1] a distance of around 230 km, the emperor decided to divide his forces because, according to the sources, he was under the impression that the Turks and Daylami defenders of Manzikert would be easily overcome.[2] Matthew of Edessa reports that Diogenes also ordered the dispatch of some ten or twelve thousand troops to the "Abkhazes,"[3] which is a rather dubious report in my view, although we should not exclude the possibility of sending smaller detachments to Georgia to procure supplies.[4] Other sources also account that the emperor dispatched separate detachments of troops to march to the strategic fortress-city of Khliat that controlled the western approaches to the region. These detachments comprised of "Uze mercenaries and the Franks under Rouselios...[sent] towards Khliat to forage for provisions," and because the emperor "considered them [Turkish defenders] of no great importance since they were few in number, he detached another not insignificant portion of his army [to march to Khliat] and placed it under the command of the *magistros* Joseph Trachaneiotes."[5]

There are two crucial points we need to emphasize here: first, that the detachment under the *magistros* Trachaneiotes "consisted of select troops,[6] difficult to withstand,

1 Nicolle, *Manzikert*, 53; Hewsen and Salvatico, *Armenia: A Historical Atlas*, 117 (Map 93).

2 "He [Romanos] regarded the enemy garrison in Mantzikert with contempt and as insufficient to withstand his onslaught." Attaleiates, *History*, 273.

3 Matthew of Edessa, *Chronicle*, 134; Smbat Sparapet, *Chronicle*, 37.

4 Leveniotes, *Η πολιτική κατάρρευση του Βυζαντίου*, 169.

5 Skylitzes, *Continuation*, 115; Attaleiates, *History*, 271–73.

6 On the (probably Armenian) identity of these troops. Antonios Vratimos, "Joseph Tarchaneiotes and the Battle of Mantzikert (AD 1071)," *Al-Masāq* 32 (2020): 156–68 at 160–63.

taking the initiative in close combat and other types of battle and ready to face danger, and far more numerous than the soldiers retained by the emperor."[7] Second, the emperor's plan was to take Manzikert with the units he kept under his command, but he did not plan to join forces with Roussel's and Trachaneiotes's detachments unless there was a dire emergency and, certainly, had not planned to march against Khliat before "Manzikert was restored to Roman rule."[8] I believe that our sources make it clear that the emperor was planning for a swift campaign to capture Manzikert and Khliat, and, perhaps, even subdue the remaining fortress-towns on the northern shore of Lake Van, replace their garrisons with imperial troops, and head back home before there was any reaction from the Seljuk leadership. Romanos was in for a shock!

Therefore, we need to return to the decisive issue of Byzantine intelligence. Attaleiates reports that "spies had already reported to him [Romanos] that the sultan was in a hurry to return to Persia."[9] This is coupled with the report dispatched by Leon Dabatenos while the emperor was probably still in Theodosiopolis in July, according to which "the sultan, it read, learning about the campaign [of Romanos to Vaasparakania] and afraid of the power [of Romanos's army], left Persia for Babylon [Baghdad]."[10] Both reports would certainly have boosted the emperor's confidence and conviction that, following his strategic withdrawal from Syria, Alp Arslan would adopt a defensive strategy in Iraq and would not attempt to march against him in Armenia.

Other sources of information proved equally misleading, as when the Armenian officer Nikephoros Basilakes arrived at the head of substantial reinforcements from Syria and Armenia and, wishing to "flatter the emperor, gave no wise/well-thought answers to his [emperor] questions."[11] There are parallels with twentieth-century intelligence officers "feeding" leaders with intelligence that they "wished to hear," but that did not correspond to the reality on the ground, thus leading to disastrous strategic decisions.[12] Finally, we have Matthew of Edessa's account of the "perfidious Romans from Diogenes's army" who allegedly dispatched a letter to Alp Arslan that read as follows: "Do not flee, for the greater part of our forces is with you."[13] Although these would probably not have been the true contents of the letter, if there ever was a letter, rumours of traitors and spies infiltrating its ranks would have posed a real challenge to any army operating under these circumstances, and we would never know the true extend of the intelligence "leak" to the enemy in these critical weeks before the major clash between the two armies.

---

7 Skylitzes, *Continuation*, 115; Attaleiates, *History*, 273.

8 Skylitzes, *Continuation*, 115; Attaleiates, *History*, 273–75; Nicolle, *Manzikert*, 51.

9 Attaleiates, *History*, 275.

10 Bryennios, *Ὕλη Ιστορίας*, 83.

11 Bryennios, *Ὕλη Ιστορίας*, 81.

12 On Stalin's strategic miscalculations and the role of Soviet intelligence for the 1939–1940 "Winter War" between the Soviet Union and Finland: Kimmo Rentola, "Intelligence and Stalin's Two Crucial Decisions in the Winter War, 1939–40," *The International History Review* 35 (2013): 1089–1112.

13 Matthew of Edessa, *Chronicle*, 133–34.

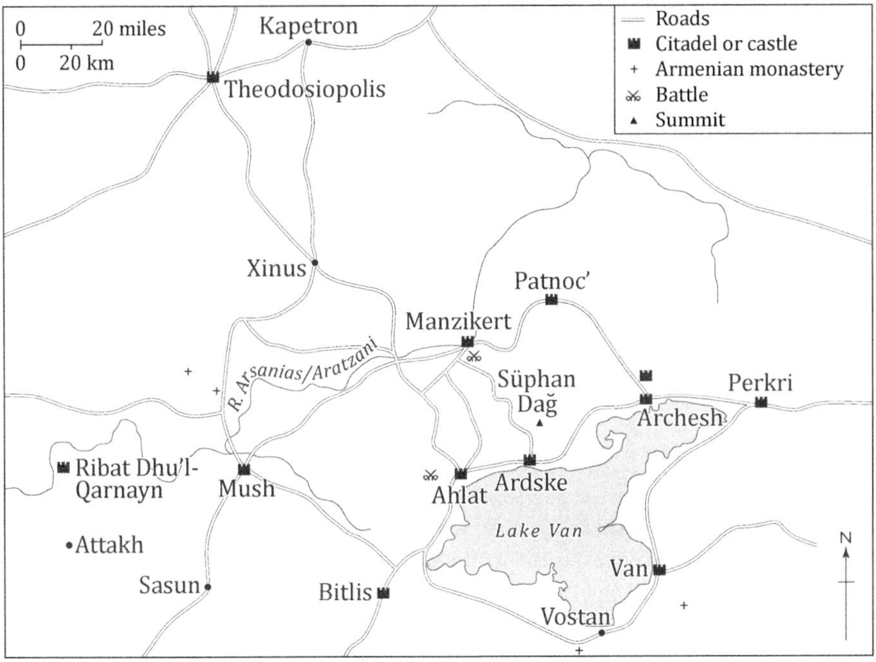

Map 2: The region of Manzikert. Map by Cath D'Alton.

The Armies Make Contact

The course of Alp Arslan's army from Khoy to Lake Van remains another mystery, and there is no consensus among modern historians as to the exact route that the sultan took. Leveniotes argues that "Alp Arslan approached the area of Manzikert from the south and not from the northern shores of Lake Van...[because] access in this way was much shorter, while at the same time he could count on the cover that the fortresses of Khliat and Artchesh, which had Turkish garrisons and controlled the northwestern shores of Lake Van."[14] I appreciate the argument of having one's flanks covered, but the approach from the southern shore of Lake Van is certainly not shorter than the one in the north, considering that the main target of Alp Arslan was the main Byzantine army under Romanos that was targeting Manzikert, rather than Khliat. Leveniotes also adds that "if the sultan's forces approached the Manzikert area from the east, as many researchers believe, they would have been forced to bypass the huge mountain range of the Süphan Daği (4,434 metres high), which was located in the east of the aforementioned forts [Khliat and Artchesh]." Nonetheless, Süphan Daği is not an insurmountable obstacle and could have been bypassed (relatively easily) either from the north, as I believe to be the case, or from the south, especially considering that its slope is fairly

14 Leveniotes, *Η πολιτική κατάρρευση του Βυζαντίου*, 172.

gentle on all sides except the north.[15] The distance between its southern slope and the coast of Lake Van (currently) stands at around seven kilometres. The mountainous terrain on the southern coast of Lake Van would have presented a much more formidable obstacle for an army on horseback, as I explained in a previous chapter.

Therefore, I find myself in agreement with David Nicolle's suggestion that Alp Arslan "could have gone along the southern shore of Lake Van, thus threatening Roussel and Trachaniotes from the rear. But such a route would be longer, more difficult, slower and would have pushed the defeated Byzantine commanders towards their Emperor, whereas in fact they fled in the opposite direction."[16] The last point is a shrewd observation by Nicolle and, based on the intelligence that the sultan would have had about the dispatch of a substantial part of the imperial army to Khliat, an ingenious strategic move that would have prevented the divided imperial detachments from re-uniting. For this reason, a Turkish advance from the northeast, coupled with the dispatch of outriders to disrupt communications between the emperor and his units in Khliat, proved highly effective, as we will see further down.

It is likely that the Turkish garrison in Manzikert would have spotted the Byzantine detachments heading to Khliat and, first, would have sent a rider to alert their comrades there. They then would have have kept intelligence lines open with the main Seljuk army which was marching to Lake Van from the east. As I will explain in more detail further down, the rugged terrain around Manzikert favours the defender, therefore any small reconnaissance parties would have had little trouble slipping past the Byzantine detachments heading to Khliat. Probably this is why Alp Arslan was able to send a substantial force of several thousand horsemen under the experienced and battle-hardened Sunduq al-Turki to reinforce Khliat which, according to Bryennios, "was defended by a fairly strong Turkish garrison."[17] In fact, we know that Sunduq al-Turki had been distinguished in the previous years for his actions in Syria and it is unlikely that he would have recalled to Iran; instead, he would probably have been ordered by the sultan to force-march his troops to Khliat through Bitlis and the west coast of Lake Van.[18] Again, it seems to me perfectly reasonable that Alp Arslan would have been able to keep his communication lines open with both his garrisons in Manzikert and Khliat and, provided he had fast and accurate intelligence, he would have been able to dispatch significant reinforcement wherever was necessary.

None of the primary sources provide any details about the first clash of the campaign between the detachments sent by Diogenes and the defenders of Khliat under Sunduq. Only a handful of our Muslim sources account for the outcome of the clash that was in favour of the Turkish defenders, reporting that "he [Sunduq] killed many of them and he led their leader away in fetters, distressed and captured," not missing the opportu-

---

15 Sinclair, *Eastern Turkey*, 1:272–73.

16 Nicolle, *Manzikert*, 54.

17 Bryennios, *Ύλη Ιστορίας*, 83.

18 al-ʿAdīm, in Hillenbrand, *Turkish Myth*, 75; Turan, *Selçuklular Tarihi*, 179; Sevim, *Anadolu'nun Fethi*, 58.

nity to add that "the sultan ordered that his nose should be cut off and his death should be postponed."[19] The reported size of the Byzantine unit defeated by Sunduq varies between al-Athīr's ten thousand to al-Bundārī's twenty thousand "stalwarts in armour," to the utterly absurd report of thirty thousand by Matthew of Edessa.[20] According to the same sources, the clash occurred on 4 Dhu'l-Qa'da 463 (Wednesday, August 3, 1071),[21] a date that should be viewed with some caution as we will see further down. However, we have no evidence of the casualties sustained by the units under Trachaneiotes and Roussel, and in what state they were in following this setback; would they have been able to complete an orderly retreat to Manzikert, if they recalled by the emperor, or were these units in disarray?

Another critical detail that could shed some light on the sequence of the events has to do with the identification of the army that the defenders of Khliat defeated on 4 Dhu'l-Qa'da 463. Attaleiates and Skylitzes make it clear that the emperor first dispatched to Khliat the Uzes and Frankish units under Roussel to pillage and to forage for supplies, and only sometime later, probably when he arrived at the vicinity of Manzikert and assessed the strength of the Turkish garrison of the town as "of no great importance," did Romanos assign his elite troops under Trachaneiotes to make their way to Khliat "to render assistance to the Uzes and Franks and the entire mercenary contingent, since the emperor had received word that an army numbering in the thousands was bearing down upon them."[22] I would imagine that the "Uzes and Franks" did not anticipate running into such a staunch resistance by several thousands of enemy horsemen called as reinforcements to defend Khliat, and once the emperor had assessed that the situation in Manzikert was manageable and under control, he dispatched Trachaneiotes to Khliat to reinforce his beleaguered units and to gather supplies. This is an indication that communications between the two armies were still open and Diogenes anticipated that "If there were some pressing emergency, he could recall them [Trachaneiotes's troops] quickly."[23]

But which detachment was defeated by Sunduq's army on 4 Dhu'l-Qa'da 463? Al-Bundārī and al-ʿAdīm report that it was only after the aforementioned clash that "the army of Byzantium caught up [with the advance guard] and they encamped at Akhlat [Khliat], besieging it," all the while "the petty king of Byzantium attacked Manzikert with his Christian helpers."[24] Contrary to Leveniotes's argument that Sunduq's forces first defeated the united detachments of Roussel and Trachaneiotes, and "the army of Byzantium that caught up" with these units at a later stage was another detachment

19 Quotations from: al-Bundārī, in Hillenbrand, *Turkish Myth*, 59. See also: al-ʿAdīm, in Hillenbrand, *Turkish Myth*, 75; al-Athīr, in Hillenbrand, *Turkish Myth*, 64; al-Jawzī, in Hillenbrand, *Turkish Myth*, 37; Sibṭ Ibn al-Jawzī, in Hillenbrand, *Turkish Myth*, 68.

20 Matthew of Edessa, *Chronicle*, 134.

21 August 3, 1071 was a Wednesday. See: al-Bundārī, in Hillenbrand, *Turkish Myth*, 59; al-ʿAdīm, in Hillenbrand, *Turkish Myth*, 75.

22 Skylitzes, *Continuation*, 115; Attaleiates, *History*, 271–75.

23 Skylitzes, *Continuation*, 115.

24 al-Bundārī, in Hillenbrand, *Turkish Myth*, 59; al-ʿAdīm, in Hillenbrand, *Turkish Myth*, 75.

under Nikephoros Bryennios "the Elder",[25] I believe that neither Bryennios's or Basilakes's units were involved in a clash with troops from Khliat, but instead experienced sustained attacks by well-trained troops from Alp Arslan's main army that had already "flooded" the region. Rather, I would argue that Sunduq's forces first defeated the Uzes and the Franks under Roussel, before Trachaneiotes's reinforcements could reach them in time to prevent this setback.

To keep track of the events in two operational theatres (i.e., in Manzikert and in Khliat) we have to remember that Trachaneiotes's elite detachment was dispatched to Khliat only *after* the emperor was informed of the unexpected resistance by thousands of enemy troops there (under Sunduq al-Turki), but *before* he set up his camp outside Manzikert to begin the siege of the city—this is crystal clear in the *History* of Attaleiates who is, by far, our most detailed source for these events.[26] I will come back to the alleged treachery of Roussel and Trachaneiotes a little later but, at the moment, I will add another comment by al-Bundārī and al-ʿAdīm who report that "The Byzantine army caught up and laid siege to Akhlat [Khliat], whilst the inhabitants were trusting in God who did not cease to help His religion," and while this was taking place, "The petty king of Byzantium attacked Manzikert with his Christian helpers."[27] I have explained that the Byzantine "army" indicated here was, in all likelihood, Trachaneiotes's detachment that, following the defeat of the Uzes and Roussel's units, decided to lay siege to Khliat after pushing Sunduq's troops back into the city. Therefore, according to these reports, both sieges (of Khliat and Manzikert) would have begun shortly after 4 Dhu'l-Qa'da 463 (Wednesday, August 3, 1071).

What exactly happened in Khliat remains a mystery as the attention of all the sources now inevitably shift to the siege of the city and citadel of Manzikert by the army led by the emperor himself. According to Attaleiates, "as the emperor approached Mantzikert, he ordered a fortified camp and palisade set up in the usual manner[28] with all their equipment someplace nearby,"[29] perhaps in one of the hills that overlook Manzikert south of the city in an attempt to block the approaches from the south (i.e., from Khliat).

Nicolle has suggested a small hill around two kilometres to the southwest of Manzikert castle as a potential place for the imperial camp, close to the modern tarmac road that links Manzikert to Khliat/Ahlat (Ahlat-Malazgirt Yolu),[30] and not far from the Çirçir and Çikrik streams that flew (until the nineteenth century) near the northern

---

25 Georgios Leveniotes, *Το στασιαστικό κίνημα του Νορμανδού Ουρσελίου (Ursel de Bailleul) στη Μικρά Ασία (1073–1076)* [The Rebellious Movement of Ursel de Bailleul in Asia Minor (1073–1076)] (Thessaloniki: Vanias, 2004), 100–101.

26 Attaleiates, *History*, 273–75.

27 al-Bundārī, in Hillenbrand, *Turkish Myth*, 59; al-ʿAdīm, in Hillenbrand, *Turkish Myth*, 75.

28 *The Taktika of Leo VI*, ed. and trans. George T. Dennis (Washington DC: Harvard University Press, 2010), Constitution XI, 195–213.

29 Attaleiates, *History*, 275.

30 Nicolle, *Manzikert*, 57. See the Google Maps "plus code": 4GPH+RRG Malazgirt, Muş, Turkey (39°08'13.5"N 42°31'46.5"E). This possible location for the Byzantine camp is also suggested

Figure 6: The double walls of Manzikert. Photo: G. Theotokis.

and southern sides of the city's walls respectively.[31] Diogenes's arrogance and vanity is again pronounced by al-Ḥusaynī, who describes the imperial camp as follows, "The Byzantine emperor erected a marquee of red satin and a canopy like it and tents of silk brocades. He sat on a throne of gold; above him was a golden cross studded with priceless jewels and in front of him was a great throng of monks and priests reciting the Gospel."[32] Right before the aforementioned description, al-Ḥusaynī is also highly critical of Diogenes's action of erecting a fortified camp for his army, because "in spite of their great number is [seen as] a sign of cowardice and failure."

Romanos arrived outside Manzikert and established a fortified campand he rode around the town to inspect its fortifications.[33] Siege machines were erected also. Attaleiates and Skylitzes suggest that, after being at the receiving end of the imperial army's powerful siege machines, the Turkish garrison only made a show of resistance and, eventually "The Armenian infantry attacked the wall outside the citadel and after many attempts took it in one sudden assault around sunset."[34] Skylitzes's account is more

---

by Adnan Çevik, "Türklere Anadolu'nun Kapilarini Açan Savaş," *Aktüel Arkeoloji* 83 (November–December 2021): 30–31.

31 Kemalettin Köroğlu and Mehmet Sait Sütçü, "Malazgirt: Bir Ortaçağ Garnizonu ve Surlari," *Aktüel Arkeoloji* 83 (November-December 2021): 82–83.

32 al-Ḥusaynī, 55.

33 On the medieval fortifications of Manzikert: Köroğlu and Sütçü, "Malazgirt: Bir Ortaçağ Garnizonu ve Surlari," 80–89.

34 Attaleiates, *History*, 277.

laconic, reporting that "When the emperor reached Mantzikert, the Turks, dumbfounded and stricken with fear at his arrival, sought and received terms."[35] Both accounts read like a general's dream in having the siege of a city come to a successful conclusion in less than a day, whereupon the garrison in the citadel sent representatives to the emperor to ask for and receive clemency.

Shortly after, however, Attaleiates reports that the garrison of the citadel refused to admit the imperial officer who was dispatched to take control of the castle, perhaps suggesting that the negotiations between the parties may have reached an impasse, adding that "they [Turks] seemed to be disregarding the treaty and refusing to honour it." But a show of force seems to have been enough to change their minds, "the emperor quickly had the battle trumpet sounded and with all his forces marched out of the camp and attacked the walls. The Turks were so astonished that they started making excuses." Yet, it seems that the emperor was willing to provide very lenient terms of surrender, for he was obviously in a hurry to take the citadel and proceed to Khliat, thus "they [Turkish garrison] requested stronger assurances for their own safety and, on receiving them, departed from the city with all their goods and bent their knee to the emperor. Their hands were not empty, however, as they did this, and they all had swords in them."[36] Following the surrender of the citadel, Diogenes arranged for sufficient provisions to be stocked for the defenders of the city before preparing to depart for Khliat.

But how far can we trust the chronology and sequence of events of these two accounts, especially if we consider that Bryennios does not even mention the siege and surrender of the city to the imperial forces? If (and this is a big *if*) we accept 4 Dhu'l-Qa'da 463 (Wednesday, August 3, 1071) to be the date shortly after which the emperor launched his siege of Manzikert, then how can we reconcile the following evidence? According to al-Bundārī and al-ʿAdīm, the surrender of Manzikert took place early "on Wednesday" (it is not clear which Wednesday of the month of August) and, crucially, that coincided with the arrival of the main Seljuk army in the vicinity of the city, presumably at its campsite on the northwestern slopes of the Süphan Dağı.[37] Couple that with the evidence by Sibṭ Ibn al-Jawzī that, following his subjugation of the citadel at Manzikert "on the promise of safe-conduct," Romanos then marched "in the direction of the sultan in a place called al-Rahwa between Akhlat and Manzikert with five days remaining of Dhu'l-Qa'da (Wednesday 25 Dhu'l-Qa'da, 463/August 24, 1071)."[38] Assuming that the Muslim sources are right in maintaining that the city's garrison formally surrendered on the promise of safe conduct on a Wednesday, this would probably have been Wednesday 25 Dhu'l-Qa'da/August 24, or perhaps the previous day that was a Tuesday, August 23, which also confirms al-Bundārī's statement that the population of the lower city of Manzikert spent the night of Tuesday-Wednesday on the town's "pave-

---

35 Skylitzes, *Continuation*, 117.

36 Attaleiates, *History*, 277–79.

37 al-Bundārī, in Hillenbrand, *Turkish Myth*, 59–60; al-ʿAdīm, in Hillenbrand, *Turkish Myth*, 75.

38 Sibt b. al-Jawzi, in Hillenbrand, *Turkish Myth*, 68.

ment" under the emperor's protection.[39] But to return to the earlier question about the discrepancy in the chronology of the events: what happened between the 4 Dhu'l-Qa'da/ August 3 and the 25 Dhu'l-Qa'da/August 24 that the sources are not telling us? I believe that the siege operation and/or Diogenes's negotiations with the garrison of the citadel probably bogged down for two to three weeks, something that the Greek pro-Romanos sources would have been keen to muffle, considering the emperor had greatly underestimated the strength and resilience of the garrison to resist him, and, instead, they give the impression that the entire operation was successfully concluded in a day or so, which I find implausible.

Whatever the case, the emperor installed a well-stocked garrison at Manzikert and made ready his army for the march to Khliat the next day, Wednesday, August 24, the day that, according to the Muslim sources, coincided with the arrival of the main Seljuk army in the region. While the imperial army was getting ready to break camp and march south, with the Greek sources describing Diogenes's efforts to improve discipline within his army's ranks by inflicting a characteristically severe punishment on a soldier who stole a donkey from a Turk (he had his nose cut off), we see the beginning of the intelligence break down that quickly led to huge confusion as to the size and the whereabouts of the enemy force(s). But which intelligence channels failed the emperor and his advisors at this critical moment of the campaign?

Intelligence has always been a fundamental aspect of warfare and by the term,[40] according to Clausewitz, "we mean every sort of information about the enemy and his country—the basis, in short, of our own plans and operations."[41] Intelligence about the enemy could be procured in two ways: by reconnaissance, when a commander openly sent out scouts to observe the enemy army and collect information about its numbers, composition and their general's intentions, or by espionage, when disguised or hidden agents operated in secret in enemy territory collecting information about the enemy. The declaration of war or the culmination of hostilities is the key point to determining where exactly espionage ends and reconnaissance begins,[42] as information could then have been gathered openly rather than in secrecy, although agents still risked their lives behind enemy lines.

On the significance of accurate intelligence gathering at a pre-war level, or before the launching of a campaign, the Byzantines had a short compilation called *On Imperial Expeditions*, addressed to Emperor Constantine VII's son Romanos. This work devises the questions that should have been asked by a commander before the launching of an expedition; there are questions about routes towards enemy lands, the geography and

39 al-Bundārī, in Hillenbrand, *Turkish Myth*, 60. The archaeological team excavating the site believe that Manzikert surrendered on August, 22. Çevik, "*Türklere Anadolu'nun*," 24–25.

40 On the methods of exchange of information and intelligence between the Byzantines and the Muslims of the eastern Mediterranean in the High Middle Ages: Theotokis, *Byzantine Military Tactics*, chapters 4 and 5.

41 Clausewitz, *On War*, 117.

42 John Keegan, *Intelligence in War: Knowledge of the Enemy from Napoleon to Al-Qaeda* (New York: Pimlico, 2003), 384.

topography of the neighbouring regions, and the detailed descriptions of plains, valleys, rivers, bridges, roads, and mountain ranges that could impede or facilitate the friendly or enemy movement of troops—and their supply of food, water, and armament. There are also questions concerning cities, castles, and other strongholds that could be besieged or avoided altogether as well as the political situation and socio-religious demography of the targeted region.[43] Similar questions are asked in the *Siyāsatnāmeh* (the *Book of Government*), the most famous work by Nizam al-Mulk, the founder of Nizamiyyah schools in medieval Persia and vizier to the Seljuk sultans Alp Arslan and Malik Shah.[44]

The sentries posted all along the frontiers played an essential role in the relay of information to the local officers and eventually to the authorities in the capital. They operated in small teams of around ten men and were responsible for patrolling a specific section of the border road network and terrain. Leo VI, author "On Skirmishing" identifies them as *viglatores* ("βιγλάτορες," Lat. "vigilator") and *caminoviglatores* ("καμινοβιγλάτορες," Lat. "caminus," meaning "the path," and "βιγλάτορες").[45] Once an invading force was in hostile territory, it was expected that special units would precede the main army and baggage train to reconnoitre the ground and local road network, and locate a suitable site for the army's encampment. The *skoulkatores* ("σκουλκάτορες") responsible for this task were watchmen and scouts, who formed the vanguard of the main army and were dispatched to "learn what is going on with the enemy and inform us."[46] Other scouts, also operating in the capacity of guides, were the *minsouratores* ("μινσουράτορες"), trustworthy, and experienced men who "were sent to spy/reconnoitre the flat regions [ahead of the army]."[47] Finally, the *prokoursatores* ("προκουρσάτορες," Lat. "precursor") was a reconnaissance unit of lightly armed cavalry numbering five hundred men marching immediately ahead of the main body of the army and whose tactical aim was to "seek contact with the enemy and set ambushes if they can, so that if the enemy is advancing in disorder without proper reconnaissance they can intercept them and strike against their *prokoursatores* to cause them panic."[48] Although none of our primary sources mention the dispatch of any such units, it seems quite remarkable that an experienced army commander like Romanos would have been as careless as to neglect sending out any *skoulkatores* or *minsouratores* to reconnoitre the ground between Manzikert and Khliat. Nonetheless, one or more of these units eventually failed, with devastating consequences for the campaign and the upcoming battle.

---

43 Constantine Porphyrogenitus, *Three Treatises*, 82–93. See also: "On Strategy," Dennis (ed.), *Three Byzantine Military Treatises*, 123–25.

44 Nizam al-Mulk, *The Book of Government or Rules for Kings: The Siyar al-Muluk of Siyasat-nama of Nizam al-Mulk*, trans. H. Darke (London: Routledge, 2012), 95–96.

45 "On Skirmishing," Dennis (ed.), *Three Byzantine Military Treatises*, 152; *The Taktika of Leo VI*, cons. 12, 270.

46 *The Taktika of Leo VI*, cons. 4, para. 26, 54; cons. 12, para. 42, 240; cons. 12, para. 97, 270; cons. 17, para. 78, 426.

47 *The Taktika of Leo VI*, cons. 4, para. 24, 54; *Praecepta Militaria*, bk. 5: 12–16, 52.

48 *Praecepta Militaria*, bk. 2:16–28, 22–24.

Wednesday, August 24

Disturbing news began to arrive in the imperial camp early on Wednesday morning, with Attaleiates reporting that "enemy forces from somewhere were attacking the soldiers' servants who were out gathering the loot, harassing and wearing them out."[49] As further reports came in, the emperor summoned Basilakes to ask his opinion on the unfolding crisis. We saw earlier that Basilakes had failed to provide the emperor with any substantial or accurate intelligence about the whereabouts of Alp Arslan and his army after he joined the campaign at the head of substantial reinforcements from Syria and Armenia. He failed the emperor once more by arguing "with his usual recklessness that they [raiders] were coming from Khliat to loot."[50] Interestingly, it is Bryennios "the Younger" who reports Diogenes's summon of Basilakes, with Attaleiates and Skylitzes accounting only that "the emperor thought that one of the sultan's officers had come up with a small detachment," neglecting to mention any council with the Armenian officer.[51]

Romanos's reaction was to take the initiative and dispatch a unit under Nikephoros Bryennios "the Elder" to intercept and repel the raiders. We have no evidence of the size or composition of this unit except that it was of "sufficient force." However, it proved to be completely inadequate to reverse the situation because, as they began to be attacked by the Turks, who released their typical "arrow-showers" on horseback from a distance, Bryennios, in turn, found himself lured into ambush and in danger of being encircled by the enemy. He decided to withdraw back to the camp and request reinforcements from the emperor. Critically, our Greek sources notice that "in this confusing mêlée...these Turks were more courageous than the others we had experienced as they charged more boldly and stood up to their opponents in hand-to-hand combat."[52] Finally, it dawned on them that they were not up against undisciplined bands of Turkmen. When the emperor dispatched Nikephoros Basilakes with a larger detachment of "local soldiers" (i.e., Armenians) to repel the Turkish forces and secure supplies for the main army, the commander acted impetuously and allowed his unit to pursue the enemy too far. After being lured into an ambush and being cut off from any reinforcements, Basilakes and many of his men were captured while the rest fell back to the camp in disarray.[53]

By Wednesday afternoon, the main Seljuk army would have properly set up camp in the northwestern foothills of the Süphan Dağl,[54] as the shortest route coming from Artchesh to Manzikert is via the northern slopes of the mountain range, dominating but unseen (this is critical!) from the plateau known as *al-Rahwa* or *al-Zuhra*.[55] We saw in

49 Attaleiates, *History*, 281.

50 Bryennios, *Ὕλη Ιστορίας*, 83.

51 Attaleiates, *History*, 281; Skylitzes, *Continuation*, 117.

52 Attaleiates, *History*, 281; Skylitzes, *Continuation*, 117.

53 Attaleiates, *History*, 283; Skylitzes, *Continuation*, 117–19; Zonaras, *Επιτομή*, 213.

54 Çevik, "Türklere Anadolu'nun," 28–31.

55 al-Jawzī, in Hillenbrand, *Turkish Myth*, 38; al-Ḥusaynī, in Hillenbrand, *Turkish Myth*, 54; al-Bundārī, in Hillenbrand, *Turkish Myth*, 60; Sibṭ Ibn al-Jawzī, 68; al-ʿAdīm, in Hillenbrand, *Turkish Myth*, 75.

a previous chapter that historians have still been unable to agree on the exact location of the plateau, yet the most convincing argument in my view is that it stretched somewhere to the south and southeast of Manzikert, between the city's walls and the (northwestern) slopes of the Süphan DağI. Following the setbacks suffered by Bryennios and Basilakes, the emperor led a substantial force out of the camp to assess the situation and identify possible enemy campsites and units operating within the vicinity of Manzikert. Attaleiates notes that Diogenes "remained on some high hills well until evening,[56] but saw no sign of an opposing force, for the Turks are wicked by nature and masters of deceit."[57] Indeed, one can feel the frustration that must have dominated the imperial council by sunset on Wednesday, August 24.

The plan was for the imperial army to march south to Khliat, but all the officers in the army would have understood that it was perilous to break camp and march south unless the route was secured. As the entire army was recalled to camp after sunset, they were in for a terrifying night as detailed in high emotion by Attaleiates, an eyewitness. He reports that the Turks attempted to impose a tight blockade of the imperial camp, as they "poured out and violently charged the Skythians [Oghuz] outside the camp and those selling army supplies."[58] What follows is a description of scenes of sheer panic inside the camp, caused by the Turkish blockade that seemed to become tighter by the hour:

> With inarticulate howling and shooting of arrows they rode around terrifying them [imperial soldiers] and placing them in a tight spot so that the victims of the attack were forced to run inside the palisade. All jammed together one after another as they were being chased and pressed into the entranceway, which caused tremendous confusion among those inside because they thought it was a full-scale attack by the enemy and that the whole camp and all their equipment would be captured. For it was a moonless night and you could not tell who was fleeing and who was pursuing, or who belonged to the other side. The Skythian mercenaries, moreover, resembled the Turks in all respects, which made the situation that night all the more confusing. It was then that a tremendous fear took over; there was talk of disaster, incoherent cries, and meaningless shouting; it was a scene of utter confusion and there was danger everywhere. Death would be, so everyone felt, preferable to what we then witnessed.

The military treatise "On Strategy" cautions the reader about the dangers of moonless night attacks by men who could "call out in the enemy's language, presenting themselves as some of their own who are fugitives and urging the others to join them."[59] Meanwhile,

56 "Do not rely only on scouts to reconnoitre the roads but the general himself, if he can do so, should, under cover, carefully observe them with his own eyes, not only the roads but, if possible, the size of the approaching enemy force and its formation. With this information he will be able to figure out what actions he can safely take against them as well as protect himself." *The Taktika of Leo VI*, cons. 20, para. 68, 561.

57 Attaleiates, *History*, 285; Skylitzes, *Continuation*, 119.

58 Attaleiates, *History*, 285.

59 "On Strategy," in Dennis (ed.), *Three Byzantine Military Treatises*, 117.

Rashid al-Din reports that Alp Arslan climbed a small hill to inspect the Byzantine camp sometime on Wednesday (morning?).[60] We should not doubt al-Din's statement that the sultan was genuinely "frightened by that army and [that] he said: 'How can we fight and repel this cruel company, with the size of army we have?'" It was one of the senior Turkmen commanders, Malik Muhammad Dānishmend, who suggested they postpone the clash for Friday, so that they have more time to prepare for battle and martyrdom. The scene was now set for the battle on Friday, August, 26.

Thursday, August 25

We read in al-ʿAdīm's account that on the following (Thursday) morning "the sultan Alp Arslan arrived with the remainder of his troops and encamped by the river, while the king of Byzantium was at a place called al-Rahwa with two hundred thousand horsemen while the sultan had fifteen thousand."[61] Setting aside the discrepancy in numbers and the absurd report about the two hundred thousand horsemen under Romanos, we see Alp Arslan moving his army closer to the imperial camp (at the *al-Rahwa* plateau) and encamping by "the river," although probably not the Arsanias/Murat River that flows to the northwest of Manzikert but one of the many smaller streams (with steep-sided beds) that intersect the apparently open ground between Manzikert and the Süphan Daği. Recent research has suggested the Çivikan Stream which flows from the Süphan Daği into the Arsanias River through Afşin.[62] This is also confirmed by Attaleiates, who reports that "the next day [Thursday] the enemy did not let up in their riding around and provoking us to fight. They also got control of access to the river which flowed nearby and tried to make the Romans suffer from thirst."[63] What Attaleiates probably means here is not the Arsanias but one of the streams mentioned earlier.

The second critical development of the day was the desertion to the Seljuks of a part of the Oghuz mercenaries in imperial service. Their number remains unknown, but sources imply that most of their comrades remained loyal to the emperor after Attaleiates himself desperately attempted to persuade them to swear oaths to that effect.[64] We can be sure, however, that this development would have contributed to the suspicion surrounding the loyalty of the Oghuz and to the general confusion in the imperial camp, while it would also have had a demoralizing effect on the rest of the troops who were still reeling from a terrifying and sleepless night.

The third critical development concerns the emperor's decision to recall the units that had been dispatched to Khliat under Roussel and Trachaneiotes. It is only Bryennios "the Younger" who reports of a military council summoned by the emperor to assess the situation and plan accordingly, while, again, he highlights the polarization within the

60 Rashid al-Din, in Hillenbrand, *Turkish Myth*, 91–92.

61 Al-ʿAdīm, in Hillenbrand, *Turkish Myth*, 75.

62 Çevik, "Türklere Anadolu'nun," 27.

63 Attaleiates, *History*, 287.

64 Attaleiates, *History*, 287–89; Skylitzes, *Continuation*, 119 (reports that these troops who defected were "under the command of Tornikios Kotertzes").

council between those who wished to go out immediately and face the main Seljuk army in a pitched battle, and those who urged caution and to await the arrival of those troops recalled from Khliat who would, of course, never arrive.[65] Attaleiates and Skylitzes fail to mention any military council summoned on Thursday morning, adding that "he [Diogenes] put off the battle. But then he gave up on receiving any help from them [units dispatched to Khliat], for he suspected that some powerful obstacle was preventing their arrival."[66] However, Romanos was still under the impression that the elite units in Khliat were simply delayed rather than been withdrawn from the region, hence the following statement of the emperor's intention that "[the emperor] decided to do battle against the enemy on the following day [Friday, August, 26] as bravely as he could with the soldiers he had at hand. He had, moreover, some hope that the others [units dispatched to Khliat] would not be delayed beyond the next day," with Attaleiates adding that "he [emperor] was unaware that their general [Trachaneiotes], upon learning of the sultan's attack against the emperor, had taken all of his soldiers and shamefully retreated through Mesopotamia until he came into Roman territory. The coward took no account of his lord or even his duty." But how fair is this assessment of the actions of Roussel and Trachaneiotes?

With the Greek sources being categorized as pro- and anti- Romanos, Attaleiates and Skylitzes were inclined to excuse the emperor's decision to dispatch almost half of his troops to Khliat. They emphasize that "the sultan arrived without any warning, and so the plan [to take Manzikert first and then march to Khliat] could not be executed," going as far as to write that "dividing his [Romanos] forces was not irrational or lacking in strategic calculation. It was fate, or rather the wrath or design of God beyond our understanding."[67] Unsurprisingly, Bryennios provides a different view, being the only one to report the decisions of the war council in Theodosiopolis that confirmed the plan to advance to Manzikert instead of the more cautious strategy of having to wait for the Seljuks to come to them; a strategy that was supported by Trachaneiotes and Nikephoros Bryennios "the Elder." We saw in a previous chapter that Bryennios "the Younger" has muffled the critical role of both Andronikos Doukas and Joseph Trachaneiotes at Manzikert, while Attaleiates fails to give any details about the actions or motivations of the magistros (or of Roussel in that matter) that day. Overall, Byzantine and Armenian sources merely accuse Roussel and Trachaneiotes of treason for fleeing west without warning their emperor of the danger he now faced.[68] But there is a surprising inference by Skylitzes in that Trachaneiotes was, indeed, privy to the plan already formulated by Andronikos Doukas to betray the emperor at the first opportune moment. He even managed to persuade Roussel to join forces with him and retreat to Roman territory

65 Bryennios, *Ὕλη Ιστορίας*, 87–89.

66 Attaleiates, *History*, 287–89; Skylitzes, *Continuation*, 119–21.

67 Attaleiates, *History*, 275; Skylitzes, *Continuation*, 117; Zonaras, *Επιτομή*, 3:213.

68 Bryennios, *Ὕλη Ιστορίας*, 83, 87–89; Attaleiates, *History*, 273, 283, 287–89; Zonaras, *Επιτομή*, 3:215; Matthew of Edessa, *Chronicle*, 134.

"through Mesopotamia" instead of rushing back to Manzikert to reinforce Diogenes, as he intended to do.[69]

There is a critical question we need to ask at this point: if the emperor dispatched messengers to Trachaneiotes and Roussel after he realized that his army was facing the main Seljuk army instead of units of irregular Turkmen, did they manage to deliver the emperor's orders for a recall back to Manzikert? This is a mystery, but we have no evidence to suggest they did, as the area between Manzikert and Khliat would have been swarming with mounted enemy troops.[70] Romanos's expectation that these units would march back from Khliat to Manzikert the next day, a distance of around 60 km, and be ready for battle was over-optimistic, to say the least.

Vratimos argues that Attaleiates's statement quoted a little earlier is sufficient evidence to assume that Trachaneiotes did indeed receive the order for a recall by Romanos, and thus chose to ignore it; let me quote the sentence again: "for he [Romanos] was unaware that their general [Trachaneiotes], upon learning of the sultan's attack against the emperor, had taken all of his soldiers and shamefully retreated through Mesopotamia until he came into Roman territory."[71] Vratimos then asks "how did the latter [Trachaneiotes] learn of the sultan's attack upon the emperor?"[72] Maybe from his own scouts who were trying to re-establish contact with the emperor and possibly warn him! Not necessarily from the emperor's messengers who, according to this theory, managed to evade the roaming Turkmen once (from Manzikert to Khliat, distance: 60 km) and deliver the recall order to Trachaneiotes and then evade them twice (Khliat to Manzikert, again, distance: 120 km) by returning to the imperial camp to report to the emperor, all the while, as I mentioned earlier, the area between Manzikert and Khliat was swarming with enemy cavalry. This would make them the swiftest (or luckiest?) messengers in Byzantine history!

Furthermore, considering Attaleiates's comment that "He [Romanos] had, moreover, some hope that the others [units dispatched to Khliat] would not be delayed beyond the next day,"[73] Antonios Vratimos argues that "the focus on Diogenes's anxious expectation for news from Khliat means that his orders had reached Trachaneiotes safely."[74] No, it does not! It is simply wishful thinking on behalf of the emperor, and a catastrophic assessment of the tactical situation. We should remind ourselves that these comments come from the pen of a pro-Romanos source who is looking for scapegoats about seven or eight years after the battle, and the same source that is ready to accuse Andronikos Doukas of treachery. I will show further down, however, that there are more credible causes for the collapse of the imperial army on Friday, August 26.

69 Skylitzes, *Continuation*, 119–21;

70 Jonathan Shepard, "Byzantinorussica," *Revue des études byzantines* 33 (1975): 220–22.

71 Attaleiates, *History*, 289.

72 Vratimos, "Joseph Tarchaneiotes," 167 (emphasis my own).

73 Attaleiates, *History*, 289.

74 Vratimos, "Joseph Tarchaneiotes," 167.

How should we make sense of the strategic reaction of Trachaneiotes and Roussel to retreat to "Mesopotamia" on the face of the rapidly deteriorating situation in the vicinity of Khliat that pitted them against strong(er) Turkish reinforcements exactly dispatched by Alp Arslan to prevent the two armies from marching back to reinforce their emperor? As David Nicolle, Daniel Beihammer and Jonathan Shepard point out, given the reputation of the two leaders, cowardice seems impossible and outright treachery also seems unlikely.[75] Following Alp Arslan's dispatchment of reinforcements to the garrison of Khliat under Sunduq al-Turki, and the subsequent defeat of the Byzantine detachment under Roussel, Trachaneiotes's army and the remnants of the first units would have launched a siege of Khliat. This is reported by our Muslim sources, yet we are at a loss as to understanding what happened after that because the focus of the sources shift to the siege of Manzikert. We can only speculate that the Byzantine commanders would have realized that they were up against a strong Seljuk army which had arrived from the south to reinforce the garrison. It is highly likely that if Alp Arslan was not in or close to Khliat, he probably headed to the hills in order to reinforce his second garrison with Turkish cavalry in Manzikert, effectively "flooding" the region between Khliat and Manzikert, consequently, cutting communications between Trachaneiotes's army and the emperor.

Assuming that this interpretation of events is correct, then the most sensible or realistic option for the beleaguered Byzantine units around Khliat would have been to retreat west along the Arsanias/Murat River valley towards (Upper) Mesopotamia (as reported by the Greek sources); first to the fortified city of Muş/Mûş, some 120 km to the west, and then to the relative safety of Melitene. This explanation may also be the reason why Romanos received no warning, as any messengers would have been easily intercepted by the Turks. Otherwise, even if Trachaneiotes did receive the recall order from Romanos—which I believe he did not, but for the sake of the argument, let's assume that he did—, and before rushing to brand him as a traitor, we should consider that the decision to withdraw west would have been based on tactical intelligence and an understanding of the situation on the ground. Perhaps Trachaneiotes considered that any attempt to reach the emperor would have been fraught with danger.

There is another intriguing suggestion in al-ʿAdīm's account on the aftermath of the Battle of Manzikert, writing that "Of the Byzantine army no one survived except the contingent which was besieging Akhlat. When the news of the defeat reached them, they left the town in terror."[76] This could very well indicate that Trachaneiotes's detachment remained in the region around Khliat and carried on with the siege of the city, as he had probably been ordered to when he was dispatched there in the first place. Does this make him a traitor for not rushing to the emperor's help?

---

75 Nicolle, *Manzikert*, 55–56; Beihammer, *Byzantium and the Emergence of Muslim-Turkish Anatolia*, 156; Shepard, "Byzantinorussica," 220–22. See also: Leveniotes, *Το στασιαστικό κίνημα*, 94–99.

76 Al-ʿAdīm, in Hillenbrand, *Turkish Myth*, 76.

Interestingly, the fact that Trachaneiotes was appointed as protoproedros and doux of Antioch to replace the pro-Romanos Chatatourios almost immediately after the emperor's blinding on June 29, 1072, may point to the Armenian's anti-Romanos or, in my view, neutral, stance during the campaign. He was considered to be a highly experienced officer and trustworthy enough by the new Doukas regime to be given a critical post far away from the centre of power.[77] I believe that branding Roussel and Trachaneiotes as traitors and designating them as scapegoats for the disaster that befell the emperor and his army on Friday, August 26, 1071, is an unfair assessment of the decisions made by the two men based on the intelligence in hand and the tactical situation on the ground.

On Thursday, there were also diplomatic overtures to the emperor by the sultan "about making a peace treaty," as reported by al-Ḥusaynī, who adds that "He [Diogenes] replied that the peace treaty would be [only] in al-Rayy. The sultan was alarmed at that."[78] Romanos's arrogance and inflexibility are also reflected in the account of al-Fāriqī, who describes a Seljuk embassy to the emperor led by the *emir* Savtekin and a senior judge from the "abbasid Caliphal court in Baghdad named Ibn al-Muhallaban:

> Romanos asked him [Ibn al-Muhallaban] about the country and its condition and he said: "Tell me: which is better, Isfahan or Hamadhan?" Ibn al-Muhallaban said: "Isfahan." Romanos said to him: "We have heard that Hamadhan is extremely cold." Ibn al-Muhallaban said: "It is so." The king said: "as for us, we will winter in Isfahan and the riding animals will be in Hamadhan." Ibn al-Muhallaban said to him: "as for the riding animals, it is true that they will winter in Hamadhan. As for you, I do not know." Then he left him and they met up to fight.[79]

Clearly, these are accounts put together to satisfy the ears of their Muslim listeners. For Hillenbrand, the "punch line" (i.e., that Romanos would not survive long after the battle known already by the audience) was bound to rouse applause and approbation.[80]

Greek sources also report the arrival of a Seljuk embassy at the Byzantine camp, "but he [Romanos] was not at all friendly in receiving them."[81] Attaleiates describes the impossible terms put forward by the emperor in that "the sultan should leave the area around the camp and settle his men further away, while the emperor could set up his own camp in that place where the Turkish forces had previously placed theirs; then they could meet to discuss a treaty." These terms would have been virtually impossible for the sultan to meet without losing face, therefore one wonders if this attitude really came as a result of the emperor's supreme confidence in his army after having already

77 Leveniotes, *Η πολιτική κατάρρευση του Βυζαντίου*, 603 (Table IZ.1.266–67).

78 Al-Ḥusaynī, in Hillenbrand, *Turkish Myth*, 54.

79 Al-Fāriqī, in Hillenbrand, *Turkish Myth*, 34. See also: Mīrkhwānd, in Hillenbrand, *Turkish Myth*, 100–101.

80 Hillenbrand, *Turkish Myth*, 126.

81 Attaleiates, *History*, 289–91; Skylitzes, *Continuation*, 121; Zonaras, *Επιτομή*, 3:215–17.

decided to take to the field or, as the Greek sources suggest, because Diogenes suspected that Alp Arslan was stalling for time, waiting for reinforcements to arrive from Iran.[82] We do know that, overnight (Thursday-Friday), the sultan and his advisers considered the emperor's peace-treaty terms and it was probably in the early hours of (Friday) morning that, according to the same sources, "without notice the emperor sounded the call to battle and for some strange reason embarked on the clash of arms. The report of this astonished the enemy when it reached them."

Friday, August 26

"He amongst you who wants to follow me, let him follow me, and he who wants to leave, let him leave my company. Here is not a sultan commanding, nor an army being commanded, for today I am only one of you and a ghāzī with you. He who follows me and gives himself to God Most High, he will gain Paradise and booty."[83] Just like their Christian counterparts, Muslim historians were very much aware of the value of adding speech to their narrative to increase the theatricality of events. It does not matter that a speech such as the one we see here by Alp Arslan, reported by Ibn al-Jawzī, is delivered in high Arabic by the tongue of a Turkish-speaking nomadic chief, and probably never took place.[84] Nonetheless, for an audience it is gripping and provides an emotional climax and religious fervour.

On that Friday morning, both sides prayed and prepared, with crosses and icons paraded before Byzantine troops. Interestingly, only Matthew of Edessa provides a basic account of this solemn process, writing that "In the morning hours the battle trumpet was sounded and heralds went forth and proclaimed the wishes of the emperor Diogenes; he promised honours, high positions, and jurisdiction over the towns and districts to all those who would courageously fight against the Persian forces."[85] Indeed, the authors of the military manuals like the *Strategikon* attributed to Emperor Maurice instruct that "suitable speeches should be given to encourage them, recalling their former victories, promising rewards from the emperor, and recompense for their loyal service to the state."[86] The same authors provide strict and accurate instructions as to the protocol that is to be followed on the day of battle itself:

> prayers should be said in camp on the actual day of battle before anyone goes out the [camp] gate. All, led by the priests, the general, and the other officers, should recite the Kyrie Eleison ("Lord have mercy") for some time in unison.

82 Attaleiates, *History*, 291; Skylitzes, *Continuation*, 121; Zonaras, *Επιτομή*, 3:215–17; Nicolle, *Manzikert*, 70; Leveniotes, *Η πολιτική κατάρρευση του Βυζαντίου*, 176.

83 Al-Jawsi, in Hillenbrand, *Turkish Myth*, 39.

84 Hillenbrand, *Turkish Myth*, 125–29.

85 Matthew of Edessa, *Chronicle*, 134.

86 Maurices' *Strategikon*, 66; *The Taktika of Leo VI*, cons. 4, 279. For more on the history of exhortative speeches in Rome and Byzantium: Theotokis and Sidiropoulos, *Byzantine Military Rhetoric*, 28–54.

> Then, in hopes of success, each meros should shout the Nobiscum Deus ("God is with us") three times as it marches out of camp.[87]

It goes without saying that soldiers are far more courageous when they believe they are facing dangers with the good will of God on their side.[88] Hence, such outpouring of religious fervour was encouraged during religious processions that took place on the days of battle.[89] On leaving the camp, however, the army was cautioned to be restrained, because "absolute silence should prevail, and no unnecessary word should be spoken. For this keeps the army in better order, and the commands of the officers are more readily understood."[90]

For decades modern archaeologists have been unable to locate the field where the Battle of Manzikert was fought, until a new project launched in 2020 aimed at finding answers. Carried out under the sponsorship of the Turkish Ministry of Culture and Tourism, and in cooperation with the Turkish General Directorate of Cultural Heritage and Museums, the Ahlat/Khliat Museum, and Alp Arslan University in Muş and Sitki Koçman University in Muğla, several teams of archaeologists and historians have conducted excavations on several sites between the modern town of Manzikert and the Süphan Daği, in the hope of finding more evidence of the clash between the two armies.

Already by the summer of 2021, archaeologists working in three different locations covering a substantial area from Manzikert to Gülkoru village, some twelve kilometres southeast of the modern town of Manzikert on the Muş-Ağri Yolu (D280), uncovered Turkish arrowheads and spearheads from the period of the battle.[91] Since the beginning of excavations, the same team of archaeologists have also unearthed 336 arrows, spears, and metal military equipment from the same period. Furthermore, it is believed that the battle took place between the Aytaç Düzü plain and the Çivi Düzü plain to the southeast of Manzikert. According to Professor Adnan Çevik from Sitki Koçman University, the excavation team conducted preliminary studies in June 2022 on a large Islamic cemetery, which is identified on historic military maps as the "cemetery of Martyrs," in the village of Afşin/Kifri, about seven kilometres southeast of the town of Manzikert—on the small road that connects the Muş-Ağri Yolu and the Ahlat-Malazgirt Yolu between the modern villages of Gülkoru and Yaramiş. Five graves were identified that contained smashed bones, showing the interred had died

87 Maurices' *Strategikon*, 33.

88 Onasander devotes an entire chapter in his General on "Taking the Omens before Battle," *Aeneas Tacticus, Asclepiodotus, Onasander*, trans. Illinois Greek Club (Cambridge MA: Harvard University Press, 1923), 429–30.

89 Karaple, *Κατευόδωσις στρατού*, 67–84.

90 Maurice's *Strategikon*, 34; *The Taktika of Leo VI*, cons. 12, chap. 53, 247.

91 A. Oğuzhan Karaçetin, "Malazgirt'te Savaş Alani Arkeolojisi," *Aktüel Arkeoloji* 83 (November–December 2021): 35–45; *Daily Sabah* (August 11, 2021), https://www.dailysabah.com/turkey/weapons-find-points-to-seljuk-byzantine-battle-site-in-turkey/news, accessed September 22, 2022.

Figure 7: One of the many smaller streams with steep-sided beds that intersect the battlefield between Manzikert and the Süphan Dağı. Photo: D. Hendrix. Used with permission.

in battle.[92] I had the great pleasure to be shown by Mehmet Sait Sütcü from Sitki Koçman University around a newly discovered group of Muslim/Turkish burials on the outskirts of the aforementioned village.[93]

With the archaeological excavations in the region still ongoing, I believe it is reasonable to assume that the battlefield lies somewhere in the area covered by the "triangle" of Manzikert with the Gülkoru (around 12 km southeast of Manzikert) and Yaramiş (around 13 km southeast of Manzikert). This is rough, rocky terrain, intersected by small streams with steep-sided beds that cut the deceptively open ground from Manzikert to the Süphan Daği. The Çivikan stream mentioned earlier is the biggest one, while to the north of Manzikert there is the Badişan stream that flows towards the town of Patnos to the east. With the ground rising gradually from 1600 metres (on the southern outskirts of Manzikert), to 1700 m (to the south of Gülkoru), and up to 1800 m at Yaramiş, the archaeologists have suggested that the most likely location for the battlefield is between these two streams.[94]

Two things that drew my attention to the terrain: first, the further away you march from Manzikert towards the Süphan Daği, the more broken and rock-strewn the ground becomes, to the point that one cannot help but wonder how easy it would have been for Alp Arslan to have had dozens if not hundreds of troops concealed behind the sizeable

---

92 Çevik, "Türklere Anadolu'nun," 31; *Al-Jazeera* (August 26, 2022), https://www.aljazeera.net/culture/2022/8/26/%D8%B0%D9%83%D8%B1%D9%89-%D9%85%D9%84%D8%A7%D8%B0%D9%83%D8%B1%D8%AF-%D8%A7%D9%84%D8%AA%D9%8A-%D8%BA%D9%8A%D8%B1%D8%AA-%D9%88%D8%AC%D9%87-%D8%A7%D9%84%D8%A3%D9%86%D8%A7%D8%B6%D9%88%D9%84, accessed September 22, 2022.

93 The group of burials was excavated right next to the school yard of the local elementary school, on the eastern edge of the village. See the Google Maps "plus code." 3HJV+C9F Yaramiş, Malazgirt/Muş, Turkey (39°04'51.8"N 42°35'36.3"E).

94 Çevik, "Türklere Anadolu'nun," 30.

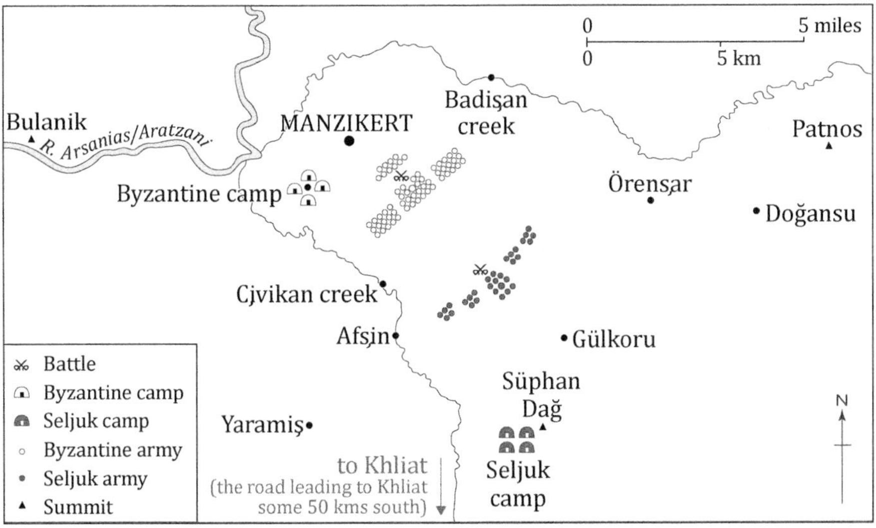

Map 3: The possible location of the battlefield and the battle formations of the two armies on the Friday morning (August 26). Map by Cath D'Alton.

rocky outcrops placed all along the battlefield.[95] And second, the rising ground explains the advantageous vantage point Alp Arslan would have had to monitor the movements of the advancing Byzantine army, while concealing the movements of his own troops. All the Greek sources confirm that either he or his scouts were quick to spot the confusion in the ranks of the imperial army that followed Romanos's order to withdraw back to the camp. The author of *The Taktika of Leo VI* warns readers accordingly, writing that "When you are in formation and have to fight against archers, avoid difficult terrain and, by all means, the lower slopes of mountains, that is, where the mountains first begin to ascend. Do not initiate hostilities there," adding that "there is a chance that your formation may be suddenly overcome by the enemy lying in ambush under cover of the high ground."[96]

Most of the Greek sources report that it was the emperor who made the first move against the Seljuks, sounding the war trumpet, *σαλπίσας το ἐνυάλιον*.[97] But it is only Bryennios who gives any details about the arrangement of the imperial army for battle. He writes that the left wing comprised mainly of the western tagmata under Nikephoros Bryennios ("the Elder"), and that the right wing included the units of Kappadokia and the rest of Asia Minor under Theodore Alyates. The emperor was in the centre and, presumably, in command of the heavy Armenian infantry and of any other palatine

95 Possible sites for Seljuk look-out posts overlooking the Byzantine camp are suggested in the map in: Çevik, "Türklere Anadolu'nun," 28–29.

96 *The Taktika of Leo VI*, cons. 14, para. 4, 291–93.

97 Attaleiates, *History*, 291; Skylitzes, *Continuation*, 121; Zonaras, *Επιτομή*, 3:215–17; Nicolle, *Manzikert*, 70; Leveniotes, *Η πολιτική κατάρρευση του Βυζαντίου*, 176.

Figure 8: The battlefield becoming increasingly rocky and broken the closer an army marched southeastwards from Manzikert towards the Süphan Dağı. Photo: D. Hendrix. Used with permission.

Figure 9: The rocky and broken terrain around where the battle may have taken place. Photo: D. Hendrix. Used with permission.

and elite mercenary units like the Nemitzoi (although we saw earlier that this unit was demoted to the rear guard) or the Varangians (if, indeed, there were any Varangians at Manzikert).[98] Finally, the emperor placed Andronikos Doukas in charge of the rear guard and in command—according to Bryennios—of "the tagma of the *Hetairiae* and that of the nobles" (*τάς τε των έταίρων τάξεις έχων και τας των άρχόντων*).[99]

Bryennios's information confirms what the military manuals describe as the basic formation for Byzantine armies from the later sixth century onwards: a tripartite line with left, centre and right divisions,[100] protected by flank guards (*άποσοβῆται*) and out flankers (*ὑπερκερασταί*) on the left and right wings respectively, both of which would have been deployed in two fifty-man *banda* of both mounted archers and lancers.[101] We may assume that the Oghuz, Patzinaks, and other Turkish auxiliaries and mercenaries would have been on the flanks of the imperial army at Manzikert, perhaps with more emphasis on the traditionally "offensive" right rather than the "defensive" left.[102] Galloping ahead of the out flankers and flank guards would have been the *prokoursatores* that we saw earlier.

Moreover, according to the author of *The Taktika of Leo VI*, "to draw up the whole army in one battle line, especially if it includes lancers, is, in our opinion, to invite a host of evils." Hence a second line, called a support line (*βοηθός*), was formed behind the first, or the *promachos* (*πρόμαχος*) line, to "discourage" any men or entire units from fleeing, to reinforce any gaps in the first line, and to lead any counterattack or to cover any retreat.[103] Moreover, we read in the *taktika* that "the lieutenant general should take his post in the centre *meros*, the other two *merarchs* in the divisions on either side, each in the centre between the *moirarchs* under his command."[104] A bit further down we also see that "the general himself, up to the moment of the charge, is to direct the formation, supervise, and adapt to the movements of the enemy. At that moment he should join his own tagma which is drawn up, not for battle, but as a sort of landmark and guide

98 For more on the role of the Armenians in Diogenes's campaign of 1071. Antonios Vratimos, "Revisiting the Role of the Armenians in the Battle of Mantzikert," *Reti Medievali Rivista* 21 (2020): 1–17.

99 Bryennios, *Ύλη Ιστορίας*, 89; Attaleiates, *History*, 287–93; Skylitzes, *Continuation*, 121–23; Zonaras, Επιτομή, 3:215–17.

100 *The Taktika of Leo VI*, cons. 12, para. 32, 235; Maurice's *Strategikon*, 16, 25, 43, 46–50. See also, Haldon, *Warfare, State and Society*, 205–8.

101 "The Praecepta Militaria of the Emperor Nikephoros II Phokas," in *Sowing the Dragon's Teeth: Byzantine Warfare in the Tenth Century*, ed. and trans. Eric McGeer (Washington DC: Harvard University Press, 1995), bk. 4:52–56, 65–67 and 69; Maurice's *Strategikon*, 49.

102 Matthew of Edessa, *Chronicle*, 135; Smbat Sparapet, *Chronicle*, 37. We read that "Foreign contingents, if stationed by themselves, should be drawn up according to their own customs. It is advantageous to employ them as assault and ambush troops." Maurice's *Strategikon*, 28.

103 *The Taktika of Leo VI*, cons. 12, para. 7–13, 219–23; cons. 12, para. 71–72, 259; cons. 18, line 10, 439. For a variation of three lines (first, second and support lines), *The Taktika of Leo VI*, cons. 18, para. 136, 489–91. See also: Maurice's *Strategikon*, 24, 50.

104 Maurice's *Strategikon*, 25.

for the first and second lines, that is, in the middle of the second line."[105] The obvious reason why the general in command was advised to stay away from the *promachos* line was that the overall command and the imposition of order and discipline (described as *τάξις* in the Byzantine manuals) in a campaigning army stemmed from its commander-in-chief. Against this background, it is easy to imagine that once the person at the top of the pyramid of command was killed or captured, the organizational and psychological consequences would be devastating: "if the leader of the whole army fails, his fall can open the way to complete disorder."[106] Romanos would not have heeded such advice.

Sometime in the mid-morning on Friday, August 26, the imperial army was ordered to leave the camp south of Manzikert and march to where Romanos and his officers would have been able to see where Alp Arslan had pitched his camp. Once again, the *taktika* caution the general that "If you are expecting combat, have your army march in formation, whether you are proceeding by *droungoi* or by *tourmai* or by entire battle lines."[107] It seems that the emperor hoped to draw the enemy quickly into a pitched battle close to Manzikert, because, as the *taktika* caution the general, "do not attempt to have the infantry formation march outside its camp for a long distance, further than two miles, so the men will not be worn out by the weight of their armament."[108] Apparently, the tactical aim of the Seljuk army would have been to deny the Byzantines what they, in turn, aimed at accomplishing, and to draw them, instead, into a meaningless pursuit into an increasingly broken terrain where their light cavalry held the advantage in mobility.

None of the primary sources for the Battle of Manzikert give any information about the battle array of the Seljuk army, with al-Bundārī only hinting at some sort of divisional arrangement on the day of the battle: "He [Alp Arslan] arranged his centre [of the army] and he divided his associates into four divisions, with each division of them being in an ambush."[109] Yet, it is reasonable to assume that Alp Arslan would probably have deployed his forces in a crescent formation of several units under their respectful leaders.[110] The sultan himself, as Mīrkhwānd implies,[111] would have taken his place at a higher ground in the rear in order to have a better view of the enemy approach, probably at the so-called "Gülkoru ridges" (the higher ground to the south of the modern village of Gülkoru) as it has been recently suggested.[112]

We saw in a previous chapter that the Byzantines were familiar with the fighting tactics of the "people from the steppes" (i.e., the Skythians, Magyars, Patzinaks, and Turks),

---

105 Maurice's *Strategikon*, 33, 43, 50.

106 Maurice's *Strategikon*, 64, 69; *The Taktika of Leo VI*, cons. 14, para. 3 and 99, 291–92 and 346–47; *A Tenth-Century Byzantine Military Manual: The Sylloge Tacticorum*, trans. Georgios Chatzelis and Jonathan Harris (Abingdon: Routledge, 2017), 28.

107 *The Taktika of Leo VI*, cons. 9, para. 5, 155.

108 *The Taktika of Leo VI*, cons.14, para. 82, 337.

109 al-Bundārī, in Hillenbrand, *Turkish Myth*, 61. Compare with: Bryennios, *Ὕλη Ἱστορίας*, 91.

110 Haldon, *Byzantine Wars*, 177.

111 Mīrkhwānd, in Hillenbrand, *Turkish Myth*, 101.

112 Çevik, "Türklere Anadolu'nun," 31.

yet it is worth looking at the recommendations of the authors of the *taktika* with regards to, first, their skills as archers; second, the feigned retreat tactics of the nomads; and, third, on how to fight against a "Turkish" crescent formation. According to the author of *The Taktika of Leo VI*, the "Turks" "are possessed by an insatiable desire for riches, they scorn their oaths and do not observe agreements they have made,"[113] a (stereotypical) view that may have added (even a little) to Romanos's decision not to agree on a peace treaty with the sultan. The authors are adamant that "they ["Turks"] devote a great deal of attention and training to archery on horseback,"[114] and because of their tribal social organization, "in battle they do not line up as do the Romans in three divisions, but in several units of irregular size, linking the divisions close to one another although separated by short distances, so that they give the impression of one battle line."[115] Finally, when it comes to the "Turkish" battle tactics, we read that "they prefer battles fought at long range, ambushes, encircling their adversaries, simulated withdrawals and wheeling about, and scattered formations."[116]

The authors of the *taktika* also give their recommendations on how to fight the "Turkish" tactics: "When it comes to battle, an infantry force in close formation opposed to their cavalry will inflict the greatest damage on them. They do not dismount from their horses and, since they have grown up riding on horseback, they do not last long on foot."[117] This surely plays on the great disadvantage of every nomadic army in history when they found themselves in close quarters with heavily armed infantry or cavalry, hence the suggestion to come to blows with a nomadic enemy as quickly as possible.[118] Further down, the author of *The Taktika of Leo VI* provides another recommendation on how to fight an "an enemy army drawn up in crescent formation," which is worth quoting in full:

> This formation, shaped like a semicircle, seems safe and unbreakable, for when the adversary falls into the hollow space of the sigma [C], then those on the flanks here and there spread out, close them in and encircle them and, in this way, fight more effectively against them. Therefore, in drawing up your line to oppose this crescent formation, divide your battle line into three sections. Make two sections, the right and the left, advance against the horns of the enemy.

---

113 *The Taktika of Leo VI*, cons. 18, para. 45, 455.

114 *The Taktika of Leo VI*, cons. 18, para. 49, 455; Maurice's *Strategikon*, 116.

115 *The Taktika of Leo VI*, cons. 18, para. 53, 457; Maurice's *Strategikon*, 117. Interestingly, Theophylaktos Simocattes (d. after 640) describes a battle between Byzantine and Avar armies near Viminacium, where the former, drawn up in three phalanxes, confronted the latter, formed of fifteen different bodies. The latter soon crowded into one long line, while the Byzantine force re-formed into a rectangular body, with the description matching the information provided in Maurice's *Strategikon*: Theophylaktos Simocattes, *Ιστορία*, ed. and trans. Despoina Tsouklidou (Athens: Kanake, 2005), bk. 8:2–3, 532–33.

116 *The Taktika of Leo VI*, cons. 18, para. 56, 457; Maurice's *Strategikon*, 117.

117 *The Taktika of Leo VI*, cons. 18, para. 61, 459; Maurice's *Strategikon*, 117.

118 Maurice's *Strategikon*, 65.

> Position the middle section so it does not advance against the hollow space but stands ready to come to the aid of the two sections moving forward. The enemy army then remains in a circle accomplishing nothing, or else it moves to attack and, by doing so, breaks up its crescent formation.[119]

The Greek sources agree on how the battle unfolded during the first half of the day, although they only provide an outline of the battle tactics deployed by both sides. The most detailed account we have comes from Bryennios, who describes the Seljuk fighting tactics almost exactly as they were detailed more than a century and a half before in *The Taktika of Leo VI*:

> So, after dividing their [Turkish] troops in many parts, he ["Taranghes"] set traps and organized entraps and gave orders to surround the Roman ranks and shoot [at them] with bows from every direction. And because the Turks were beating their horses, they [Romans] were forced to pursue them, and so they did, while they [Turks] pretended to flee; and falling into the traps and entraps, they [Romans] suffered great loss. And while the emperor, wishing to bring them to a decisive (ὁλοκληρωτική) battle, followed them slowly, waiting to meet with the Turkish army, and thus decide the outcome of the war, [but] the Turks scattered in all directions.[120]

According to Bryennios's account, Alp Arslan had assigned the bulk of his troops (there is no mention here of divisions or parts, but of troops) to his Turkoman leader called Tarang/Tarankoğlu,[121] while he, instead, stayed in the rear to have a better vantage point to monitor enemy movements. Unsurprisingly, we see the Turks employing their usual nomadic tactics of "feigned retreat" that sought to exploit their extraordinary skill with the bow and their mobility as light horse-archers, usually by staying out of reach of their opponents' weapons and using hit-and-run tactics in waves, "showering" the enemy with arrows, making extensive use of the famous "Parthian shot" as they went. When the enemy's forces were drawn out in pursuit and became exhausted, the nomads would wheel around and deal a decisive blow. However, we need to stress that Roman officers and authors regularly designated the aforementioned tactics as "Scythian" or "barbarian" for many centuries before Manzikert, where the first clear instance such tactics being employed by a Roman army refers to Aurelian's victory over the Palmyrene *cataphracti* at Immae in 272.[122]

---

119 *The Taktika of Leo VI*, cons. 20, para.184, 601.

120 Bryennios, *Ύλη Ιστορίας*, 91. See also: Attaleiates, *History*, 293; Skylitzes, *Continuation*, 121–23; Zonaras, *Επιτομή*, 3:217.

121 "Taranges," *Prosopography of the Byzantine World*, https://pbw2016.kdl.kcl.ac.uk/person/111759/, accessed January 12, 2023.

122 Zosimus, *New History*, trans. Ronald T. Ridley (Canberra: Australian Association for Byzantine Studies, 1982), bk. 1, line 50, 3–4; Philip Rance, "Battle," in *The Cambridge History of Greek and Roman Warfare Volume 2: Rome from the Late Republic to the Late Empire*, ed. Philip Sabin, Hans van Wees, and Michael Whitby (Cambridge: Cambridge University Press, 2007), 369–70.

Late Persian cavalry also appear to have employed similar tactics.[123]

Modern historians like to use "natural phenomena" to describe the massive but concentrated use of "arrow-power" by nomadic cultures in battle,[124] with an "arrow-storm"' being unleashed at between two to three hundred metres, although composite bows were significantly more effective at distances under one hundred metres.[125] We should note that horse-archers did not aim at specific targets, but rather fired their arrows at a high angle into a predetermined "killing zone," thus leaving their enemies stunned and unable to react. These "'hit-and-run" tactics, known as the caracole, can be described simply as "waves" of horse-archers firing several arrows as they charged, and then circling back to their lines after having fired their final shot roughly forty to fifty metres from the enemy line, before wheeling around to fire the "'Parthian shot," a highly skilled technique of unleashing a bow while turning at 180 degrees to the gallop of the horse (essentially, shooting backward whilst riding away).[126] The "feigned retreat" was also a typical tactic from the steppes that was practised long before the famous twenty-day retreat by the Scythians before the mighty Persian army led by Darius I in 512 BC, while there are numerous examples of it being used by the Romans and the Byzantines.[127] A smaller force charged the enemy front and then retreated, drawing the enemy in pursuit to over-extend its formation, then other forces would attack from the flanks at a pre-arranged location (often a defile), with the aforementioned token force wheeling around and attacking from the front to close the trap.

The Seljuk horsemen did not care about the cohesion of their formation, relying more on surprise raids, ambushes, and harassment tactics; they lagged in hand-to-hand fighting. This is how al-Bundārī's describes the Seljuk tactics at Manzikert:

> The petty princeling of Byzantium attacked with his troops and he snatched [his pre-ordained] fate and he approached like flowing [water] seeking a resting place and night robbing day. The horsemen of Islam held firm (thabitat), then they jumped (wathabat) and roamed freely (jalat) and they were not afraid (wajalat) and they dragged the Byzantines until the ambush was behind them and death stood in front of them.[128]

---

123 *The Chronicle of John Malalas*, trans. Elizabeth Jeffreys, Michael Jeffreys, and Roger Scott (Melbourne: Australian Association for Byzantine Studies, 1986), 18:60.

124 Actually, the use of natural phenomena to describe the ferocity of the firepower of the nomadic bows and arrows was recorded for the first time at the Battle of Mohi in 1241, fought between the Mongol Empire and the Kingdom of Hungary during the Mongol invasion of Europe.

125 Paul E. Klopsteg, *Turkish Archery and the Composite Bow* (London: Butler, 1987); David Nicolle, *Crusader Warfare Volume II: Muslims, Mongols and the Struggle against the Crusades* (London: Hambledon, 2007), 179–80.

126 For the nomadic "arrow-storm', concentrated firepower, "caracole" tactics and the "feigned retreat." Timothy May, *The Mongol Art of War: Chinggis Khan and the Mongol Military System* (Barnsley: Pen & Sword, 2007), 69–71; Morillo, Black, and Lococo, *War in World History*, chapter 6; Nicolle, *Crusader Warfare*, 2:141–43; Smail, *Crusading Warfare*, 75–83.

127 Herodotus, *The Histories*, 4:46:1–3; Theotokis, *Byzantine Military Tactics*, 220, 250, 281, 284.

128 al-Bundārī, in Hillenbrand, *Turkish Myth*, 62.

Having to face an elusive enemy is the kind of tactic that could easily drive an opponent who relied more on fighting in the fray to despair. The heavier and slower Byzantine horsemen would have attempted to chase down their lighter and faster opponents, but this would have entailed grave danger and compromised cohesion, especially in a broken and elevating terrain like the one south of Manzikert wherethe Seljuks could then quickly reverse their horses, surround the nearest unit, and crush it. It was the inability of the Byzantine forces to come to grips with the Turkish mounted archers at the Battle of Anzen/Dazimon (July 22, 838) that precipitated the dissolution of the Byzantine line and the disintegration of Emperor Theophilos's army.[129] In a twist of fate, while Theophilos was leading two thousand men to strengthen the right wing of his army before noon, he had to cross to the rear of the main body of his troops, and was thus out of sight of the men fighting in the front line, causing them to believe he had been killed, leading themto a disorderly retreat.

We read in Bryennius's account that the main Byzantine battle line advanced at a slow but steady pace, in an attempt to keep in contact with the rear guard, moving in a southeasterly direction towards the ridge where the lined-up Seljuks could be seen. It is also where their enemies responded with a devastating barrage of arrows on the advancing imperial units before withdrawing at a bowshot distance to repeat the same tactic. Since each horse archer would have been equipped with around sixty arrows in his quiver, the Turks could have maintained this barrage for more than an hour, perhaps longer if the number of men in each wave varied. In addition to the warrior's own personal quiver, extra quivers were attached to their remounts' saddles to ensure a supply of arrows.[130] Bear in mind what I mentioned earlier about the Turkmen horse-archers not aiming at specific targets, just like their English counterparts with the longbow a few centuries later. They rather fired their arrows at a high angle into a predetermined "'killing zone," thus having little impact on well-armoured men on horse or foot. To quote John Keegan: "these arrows cannot...given their terminal velocity and angle of impact, have done a great deal of harm."[131] Rather, it was the psychological impact of the clouds of arrows that fell upon the infantry and cavalry that did so: wounding men and horses, spreading confusion and panic among the troops, and breaking or disrupting their march and unit cohesion, especially since those who were at the receiving end could not counterattack to neutralize the threat.

We may also assume that, by being deployed in a crescent formation, the Seljuk "horns" would have concentrated their efforts on the left and right flanks of the imperial army, with the Turkish "centre" (the "belly" of the crescent) withdrawing some-

---

129 Haldon, *Byzantine Wars*, 82–86; Irini-Sofia Kiapidou, "Battle of Dazimon, 838," *Encyclopaedia of the Hellenic World, Asia Minor Online*, http://asiaminor.ehw.gr/Forms/fLemmaBody.aspx?lemmaid=7921, accessed February 2, 2023.

130 May, *The Mongol Art of War*, 50–53; Nicolle, *Crusader Warfare*, 2:140–41.

131 Keegan, *Face of Battle*, 70, 94–98; David Whetham, "The English Longbow: A Revolution in Technology?," in *The Hundred Years War, Part 2: Different Vistas*, ed. L. J. A. Villalon and D. J. Kagay (Leiden: Brill, 2008), 213–30. Kelly DeVries, *Medieval Military Technology* (Toronto: University of Toronto Press, 2012), 38–42.

what faster. Likewise, we saw earlier that the author of *The Taktika of Leo VI* advised the Byzantine general in a similar fashion to "advance against the horns of the enemy." However, the critical point in the latter recommendation is that the general should "position the middle section so it does not advance against the hollow space [of the enemy crescent] but stands ready to come to the aid of the two sections moving forward" as forewarning to prevent division at the centre of the tripartite line (the one led by Romanos at Manzikert) from becoming detached from its flanks and being drawn into the enemy crescent. None of our primary sources, however, can confirm this assumption.

The critical question to answer at this point is how long the Seljuk horse-archers would have been able to sustain such a concerted barrage of fire against the Byzantine line. To the same effect, how long would the advancing imperial units have been able to withstand the Turkish onslaught? Timing is of the essence here, and Muslim sources agree that Alp Arslan's counterstroke was delivered at the end of the afternoon prayer, when Muslim preachers were delivering their sermons from the minbars or pulpits.[132] This could very well have been the time when the sultan was informed by his scouts, who were concealed behind the rocky outcrops in higher ground observing enemy movements, about the "sudden misfortune that hit the Romans, [and] told the sultan what was going on, and insisted that he turn around."[133] What, exactly, was this sudden misfortune?

Sometime earlier that afternoon, Romanos issued a critical order for "the imperial banner to be turned around as a signal for the troops to return to the camp."[134] The emperor had probably realized that he had marched his lines far from his fortified camp, which was itself almost empty and vulnerable to attack, while it is likely that the gaps between his centre and wings grew longer than they should have been, as I mentioned earlier. His troops would have undoubtedly become tired, thirsty, and perhaps short of supplies as well, although we have no indication as to what happened between the mid-morning and the afternoon order by Romanos to retreat to camp. There may have been some sort of a lull around high noon, perhaps when the imperial units managed to take the Turkish camp. Although, again, this is just speculation and the sources do not confirm any of this.[135] Unfortunately for the Byzantines, however, the signal to retreat back to the camp was misinterpretedby some of the divisions in the flanks, who apparently

132 al-Ṭurṭūshī, in Hillenbrand, *Turkish Myth*, 29; al-Fāriqī, in Hillenbrand, *Turkish Myth*, 34; al-Jawzī, in Hillenbrand, *Turkish Myth*, 38; al-Ḥusaynī, in Hillenbrand, *Turkish Myth*, 55; al-Bundārī, in Hillenbrand, *Turkish Myth*, 61; Ibn al-Athīr, in Hillenbrand, *Turkish Myth*, 65; Sibṭ al-Jawzī, in Hillenbrand, *Turkish Myth*, 69; al-ʿAdīm, in Hillenbrand, *Turkish Myth*, 76; Mīrkhwānd, in Hillenbrand, *Turkish Myth*, 100.

133 Attaleiates, *History*, 295.

134 Attaleiates, *History*, 295.

135 Haldon, Leveniotes, and Nicolle write about the sack of the Turkish camp, but none of them provides any reference to the sources: Nicolle, *Manzikert*, 73; Leveniotes, *Η πολιτική κατάρρευση του Βυζαντίου*, 178; Haldon, *Byzantine Wars*, 178.

thought that the centre, led by Romanos, had been defeated and/or that the emperor had fallen.[136]

Setting aside any accusation of treachery, the fact that some of the Byzantine units proved unable to perform an orderly withdrawal back to the relative safety of the camp is a strong indication of the low morale and the lack of training and discipline in the imperial army. According to the *Taktika of Leo VI*, "Untrained men turn out to be totally ignorant and blind in the face of sudden and unexpected actions." The source adds a little later that "it is necessary for you [general] to exercize the army and to make the soldiers skilled in war by getting them used to it," arguing that "Even if they do acquire a little experience, they swiftly retreat, unable to bear the labour and the dangers."[137] It is this kind of advice that could help us understand what actually happened during the afternoon of Friday, August 26—an order to withdraw that disintegrated into a rout because of the collapse in communications and order. It was the moment that Alp Arslan had been awaiting, and his scouts were quick to spot and relay the intelligence back to him.

Another reason for Alp Arslan having to hold his attack against the Byzantines until late in the afternoon was the wind. Al-Ḥusaynī notes that "a dusty wind blew which blinded the eyes of the Muslims, and the sultan's army almost took flight...The sultan kept on repeating this entreaty and weeping until the directions of the winds were reversed and the eyes of the infidels were blinded."[138] This factor may have also played a role in the Turkish barrage being less effective than anticipated. But, as the sources assure us, there was a "miracle" in the late hours of the afternoon and the wind changed direction, a development that was interpreted as divine approval of Alp Arslan's counterattack.

We are not told which of the two flanks of the imperial army was the first to buckle, but Bryennios is the only source to report that "then they [Turks] turned back, and when they attacked with great violence and shouting against the Romans, they put the right horn to flight,"[139] which included the Kappadokian and other Anatolian units under Alyates, although our source does not mention the miscommunication with the banner signal to withdraw back to the camp. The role of the rear guard was essential under these circumstances, in that "if the first line happens to retreat, it may find refuge in the empty spaces between the divisions [of the second battle line], and with them it will seem to form one battle line."[140] Therefore, at this critical moment in the battle, as the troops of the right flank fled towards the safety of the camp, the rear guard, instead of rushing to

---

136 On the different kinds of military signals, like "voiced," "semi-voiced" and "mute." Vegetius, *Epitome of Military Science*, trans. N. P. Milner (Liverpool: Liverpool University Press, 2001), bk. 3, 5. "We do not consider it desirable that many trumpets be sounded or blown during battle, for this causes disturbance and confusion, and commands cannot be properly understood." Maurice's *Strategikon*, 33. Repeated in: *The Taktika of Leo VI*, cons. 12, para. 53, 247; Sylloge *Taktikorum*, 36.

137 *The Taktika of Leo VI*, cons. 7, para. 1–2, 105. Compare with: cons. 14, para. 97, 345.

138 Al-Ḥusaynī, in Hillenbrand, *Turkish Myth*, 55; Mīrkhwānd, in Hillenbrand, *Turkish Myth*, 101.

139 Bryennios, *Ὕλη Ιστορίας*, 91.

140 *The Taktika of Leo VI*, cons. 18, para. 136–37, 491–93.

assist, reportedly sped up its own withdrawal. This was interpreted by Attaleiates and Skylitzes as treachery by its commander, Andronikos Doukas.[141] For Nikephoros Bryennios "the Younger," the failure of the Byzantine rear guard to support the main division under the emperor enabled the Seljuks to "surround the emperor and assail him on all sides,"[142] but there is no hint of treason.

It is impossible to be sure which of the aforementioned views is closer to the truth, yet the failure of Andronikos's division to come to the aid of the collapsing front can be attributed to the great distance that may have developed between the rear guard and the main line during the hours of the afternoon and, as a result of this miscalculation, they left the promachos line without support.[143] Therefore, once Andronikos would have realized that the right flank had already dissolved and the Turks firmly tried to encircle Romanos's division, he decided to retreat. I have explained elsewhere the example of the annihilation of the unit of Varangians at the Battle of Dyrrachium, on October 18, 1081, when the emperor Alexios Komnenos, seeing the right wing of the Normans in disorderly retreat, signalled a general advance.[144] But with the centre and right divisions of the Byzantines already involved in a moderate skirmish with their opposite Norman divisions and the Varangians advancing well beyond the Byzantine line of march, the tactical mistake was Alexios's failure to catch up and support the Varangians' advance, thus leaving them cut off and annihilated by the Normans. Anna Komnene is very careful to minimize her father's blunder at Dyrrachium, arguing that "the axe-bearing barbarians [Varangians] and their leader Nabites had in their ignorance and in their ardour of battle advanced too quickly and were now a long way from the Roman lines, burning to engage battle with the equally brave Franks [Normans]."[145] We should expect the same hunt for scapegoats from the pro-Romanos sources.

Seeing what was happening on the Byzantine right wing and perhaps being informed by scouts that the enemy's rear guard was farther away than it should have been, Alp Arslan saw this as the right moment to strike. Muslim sources describe that when the time of the afternoon prayer drew nigh, "He [Alp Arslan] threw down the bow and the arrows and took up the sword and the mace," thus indicating to his troops that he was about to enter close combat. "[A]nd he tied the tail of his horse with his [own] hand," another symbolic action illustrating that he would enter the battle and wanted to make it harder for opponents to disturb his horse by grabbing hold of its tail.[146] Muslim sources also confirm that "Alp Arslan had verified [the position of] the tent, banner, horse and outward appearance of the king of Byzantium [Romanos]. Then he said to his men: 'Let

141 Attaleiates, *History*, 293–95; Skylitzes, *Continuation*, 123; Zonaras, *Επιτομή*, 3:217.

142 Bryennios, *Ύλη Ιστορίας*, 91–93.

143 "For during battle, it (second line) should not be so far behind the first line that it cannot provide support, nor should it be so close that it may get mixed up with it in battle, especially when a lot of dust is being kicked up." Maurice's *Strategikon*, 31.

144 Theotokis, *Norman Campaigns*, 161.

145 Komnena, *Alexiad*, bk. 4, 6.

146 ibn al-Athīr, in Hillenbrand, *Turkish Myth*, 65; Sibṭ al-Jawzī, in Hillenbrand, *Turkish Myth*, 69; al-ʿAdīm, in Hillenbrand, *Turkish Myth*, 76. See also: Nicolle, *Manzikert*, 76–77.

each of you not fail to do as I do and strike with his sword and shoot with his arrow where I strike with my sword and shoot with my arrow." Then all of them launched an attack as one man on the tent of the king of Byzantium."[147] That Alp Arslan focused his attack on the centre of the Byzantine main line where he could probably see the imperial banner, is also confirmed by Sibṭ Ibn al-Jawzī.[148] Once the Byzantine right wing had dissolved into a panicky retreat, the Seljuks would have found it a lot easier to encircle Romanos's main division in the centre, especially since the rear guard proved unable (or unwilling?) to dispatch any reinforces to stabilize the front. Only Bryennios reports that his relative, Nikephoros Bryennios ("the Elder"), in command of the left wing, made efforts to support the centre but found his division being attacked from the rear (νῶτα) and surrounded, so he too was forced to retreat, eventually breaking up in flight.[149]

What all Greek sources agree on is, of course, the heroic stance of Emperor Romanos throughout this critical stage of the battle.[150] Following the communications blunder with the signal to withdraw to the camp, Diogenes "took a stand with his own men, trying in the usual way to check the flight of his men, but nobody was listening to him." While his division of primarily Armenian infantry beared the brunt of Alp Arslan's attacks, he "ordered the men about him not to yield or let up in the slightest and for quite a while he valiantly defended himself," perhaps hoping to launch a counterattack that could have allowed for his soldiers to reach the camp safely. But after being wounded in the arm and having his horse shot from under him, with collapse inevitable, Attaleiates describes an emotional scene that could well be from an ancient Greek drama, involving the capture of his hero:

> The enemy surrounded the emperor but they did not find it easy to capture him quickly, for he was an experienced warrior and general who had faced many dangers. He vigorously defended himself against his attackers and killed many of them, but in the end, he was wounded on his hand by a sword. His horse, moreover, had been shot with many arrows and he was fighting on foot. Toward dusk he tired, surrendered, and—O to suffer such a thing!—was taken prisoner.[151]

147 al-Ṭurṭūshī, in Hillenbrand, *Turkish Myth*, 29.

148 Sibṭ al-Jawzī, in Hillenbrand, *Turkish Myth*, 69.

149 Bryennios, *Ὕλη Ιστορίας*, 93.

150 Attaleiates, *History*, 295; Skylitzes, *Continuation*, 123–25; Bryennios, *Ὕλη Ιστορίας*, 93; Zonaras, *Επιτομή*, 3:219.

151 Attaleiates, *History*, 297.

Chapter 7

THE AFTERMATH OF THE BATTLE OF MANZIKERT

Chaos in the Camp

Because he was present in the fortified Byzantine camp, Michael Attaleiates was able to offer a highly emotional personal account of the unfolding disaster, as he experienced it through what he saw and what he heard from the soldiers who fled from the battlefield.[1] There is a mixture of feelings in Attaleiates's account; panic and confusion about how the events unfolded, and uncertainty over who to ask for information: "Meanwhile, the flight of the others had led them to cluster outside the camp palisade and all were shouting incoherently and running about in disorder; nobody could say what exactly was going on." Critically for Attaleiates, the fate of the emperor was of utmost importance, hence we read the contradicting reports that he received throughout Friday afternoon, "some claimed that the emperor was firmly resisting with what was left of his army and that he had routed the barbarians. Others said that he had been killed or captured. Everyone had something different to report, claiming victory for each side and then alternately denying it."

Tthere was the growing awareness of the outcome of the battle, coupled with Attaleiates's desperate attempts to avert the unfolding disaster by trying to convince some of the retreating soldiers to return to their posts, like in the aftermath of the desertion of some of the Oghuz on Thursday (August 25): "Finally, many of the Kappadokians who were with the emperor, one group after another, began to desert. Whether I myself was trying to stop many of the soldiers from running away and getting them to return to their posts to save us from defeat, I leave it to others to report."

It was when Attaleiates had finally accepted that everything was lost, and a feeling of sadness and desperation set in that we read his harrowing words about the fate of the emperor and his army. There are reports of the Turks seizing and looting the imperial camp, but the loss of prestige was far harder to bear:

> It was like an earthquake with howling, grief, sudden fear, clouds of dust, and, finally, hordes of Turks riding all around us. To the extent that his speed, eagerness, and strength permitted, each man sought safety in flight. The enemy chased us, killing some, capturing some, and trampling others underfoot. It was a terribly sad sight, beyond lament and mourning. For what could be more pitiable than the entire imperial army in flight, defeated and chased by inhuman and cruel barbarians, the emperor defenceless and surrounded by armed barbarians, and the tents of the emperor, the officers, and the soldiers taken over by men of that ilk, and to see the whole Roman state overturned, and knowing that the empire itself might collapse in a moment?

1 Attaleiates, *History*, 295–97; Skylitzes, *Continuation*, 123–27.

It is easy for a victorious general to boast, "Victory comes from the Heavens," but as we all know, so does defeat! For God's "chosen People" of Romania, military success over infidels was associated with divine blessing and, conversely, military defeat with the withdrawal of blessing because of sin. There is no question in Attaleiates's mind as to who was the ultimate judge on the battlefield: "It was with such an awareness of their human fallibility and with levelheadedness that the Turks reacted to their victory, neither boasting loudly, as people tend to do when things go their way, nor ascribing the deed to their own powers, but rather they ascribed the whole thing to God, as a feat that surpassed their own power to accomplish." We will see that the Muslim sources emphasize the fact that Diogenes was allegedly captured by a lowly soldier, thus highlighting how Islam humbles the proud, whereas Attaleiates saw it as an opportunity to acknowledge the magnanimity and humility of Alp Arslan and the Turks. Where he recounts the humane treatment of Romanos by the victorious sultan, Attaleiates adds, "Thus God's will was shown here as well to be just and infallible, for not only the others but the captured emperor himself came to the opinion that the sultan was worthy of victory."[2] This is not, as it will be portrayed by Muslim sources, the triumph of Islam over Christianity and the first step in the epic story of the Turkish-led dynasties in Anatolia and the Middle-East, but another opportunity to reflect on the inexorable workings of God's will in human history.

The Captured Emperor

For Romanos, the game was up—at least against the Seljuks; we read the dialogues between the sultan and the captured emperor in the account of Ibn al-Jawzī that resonate with a strong sense of theatricality to induce satisfaction in the ears of the Muslim listeners:

> He [Romanus] said: "The sultan should say what he wants." He [Alp Arslan] said: "I want a million dinars." He [Romanus] said: "by God, you will deserve the kingdom of Byzantium, if you spare my life, but I have spent money and, since I have ruled over them [the Byzantines], I have used up from the wealth of Byzantium ten thousand dinars on renewing the army and on the wars which I have fought with it [the money] until this present battle of mine, and I have made them poor by that."[3]

Having been chosen to lead the empire's counterattack to stem the Turkish penetration of Anatolia because of his extensive military experience, this certainly was a terrible blow to Diogenes's prestige, something that he allegedly acknowledged to the sultan with a palpable sense of bitterness, "[Romanus] said to him [Alp Arslan]: 'If you show favour to me, my being sent back will be speedy before the Byzantines appoint a king

---

2 Attaleiates, *History*, 301.

3 al-Jawzī, in Hillenbrand, *Turkish Myth*, 40.

other than me.'"[4] Romanos knew he was expendable, and this was the basic challenge of imperial legitimacy in the Byzantine Empire.[5]

There is little doubt that the Battle of Manzikert would have made a good story to its Muslim audience and that the necessary ingredients to excite their interest and induce satisfaction are present (fantasy and exaggeration, drama and suspense, and emotional and physical humiliation of the anti-hero, resulting in a happy ending); the same applies to the largely fictional details about the encounter (and dialogues) between the sultan and the emperor in the aftermath of the battle. For Hillenbrand, there are three points we have to pay attention to in the narrative accounts about the famous encounter:[6] first, the fact that the emperor was captured by a lowly *ghulām* who was rejected as being unworthy of payment for military service; second, that Romanos was ritually humiliated in front of Alp Arslan before being made to offer a drink to the sultan as his cupbearer; and third, the "literary device"" of a tripartite structure in the famous dialogue between Alp Arslan and Romanos.

The story of the slave-soldier capturing the emperor has been ingrained in the Muslim literary memory of the battle and has been copied and expanded throughout the following centuries. For the Muslim audience, Romanos's pride and arrogance are punished with his capture by a lowly slave, thus becoming some sort of a figurehead for how God punishes the arrogant and the insolent, while Alp Arslan dominates the story with his humility and lack of ostentation (hero versus anti-hero). The earliest account/version of the capture of the emperor is by Ibn al-Jawzī, who reports that a senior Seljuk officer named Said al-Dawlah Gawhara'in went to Alp Arslan and said:

> "O Sultan! One of my ghulāms mentioned that the king of Byzantium is in captivity with him." This ghulām had been presented to Nizam al-Mulk with the whole army and he [Nizam al-Mulk] had looked down on him and rejected him. Overtures had been made about his case but he [Nizam al-Mulk] refused to appoint him and said scornfully: "Perhaps he will bring us the king of Byzantium as captive." So, God Most High made the imprisonment of the king of Byzantium occur by his hand.[7]

The aforementioned account becomes more elaborate around the mid-thirteenth century by Sibṭ Ibn al-Jawzī who includes an actual interrogation of the *ghulām* who captured Romanos by Alp Arslan, in an effort to add to the humiliation of the emperor by relaying the details of the episode.[8] Interestingly, the *ghulām* allegedly reported that

---

4 al-Jawzī, in Hillenbrand, *Turkish Myth*, 41.

5 Anthony Kaldellis, *The Byzantine Republic, People and Power in New Rome* (Cambridge, MA: Harvard University Press, 2015).

6 Hillenbrand, *Turkish Myth*, 126–29.

7 al-Jawzī, in Hillenbrand, *Turkish Myth*, 39; al-Ḥusaynī, in Hillenbrand, *Turkish Myth*, 56; al-Athīr, in Hillenbrand, *Turkish Myth*, 65; al-ʿAdīm, in Hillenbrand, *Turkish Myth*, 76.

8 Sibṭ Ibn al-Jawzī, in Hillenbrand, *Turkish Myth*, 69–70. Compare with: Mīrkhwānd, in Hillenbrand, *Turkish Myth*, 102–3.

when he launched his attack against the emperor, the latter was surrounded by *saqaliba* troops, a term used to describe Slavs and other fair-haired people and, perhaps, the Varangians as well.

A far more elaborate version comes from the account of the early fourteenth century by Rashīd al-Dīn Faḍlullāh Hamadānī (d. 1318/717),[9] where the slave-soldier who captures Romanos was of Byzantine (Christian?) origin an behaves duplicitously towards the emperor by offering to save him from captivity and take him back to Byzantium in the night, but eventually betrays him to Alp Arslan. Another thing that would have placed Diogenes in a bad light for a Muslim audience was the fact that he was allegedly overwhelmed by the effects of alcohol at a banquet that was organized for him by the sultan. Interestingly, the Andalusian chronicler al-Ṭurṭūshī, who penned the earliest account we have of the Battle of Manzikert and its aftermath in his *Kitāb Sirāj al-Mulūk* (completed in 1122/512), does not mention the story of the slave-soldier.

The overall theatricality of the encounter between Romanos and Alp Arslan is displayed in the sultan's treatment of the emperor as his prisoner. In particular, the ritual ceremonies enacted at this point deploy powerful symbols that aim to emphasize the emotional and physical humiliation of Romanos by a set of quasi-ritualistic indignities that Alp Arslan exacts on him before he set him free. We read in the earliest account by al-Ṭurṭūshī that:

> He [Romanos] was led with the rope round his neck and a proclamation was made about him: "Who will buy the king of Byzantium?" They went on like that going round with him to the tents and the Muslims' houses and the announcement for him was made in dirhams and fulus [i.e., small coinage]. Nobody paid anything for him until they sold him to a man for a dog...He [Alp Arslan] said: "That is just, because the dog is better than he is!"[10]

For Hillenbrand, this theme of an alleged barter transaction between Romanos and a dog reflects at a deeper level the emperor's status as a polluted creature, for the dog is considered unclean in Islam.[11] In later accounts, this humiliation is followed by another:

> The sultan ordered that his chains and collar be undone, then he said: "Give him [Romanos] a drinking goblet." So, it was given and he [Romanos] thought it was for him and so he wanted to drink from it, and he was prevented from doing so, was ordered to serve the sultan and to walk towards him and give it [the goblet] to him. So, he nodded a little towards the ground according to the Byzantine custom and came towards him [Alp Arslan]. The sultan took the goblet, pulled his hair, put his face on the ground and said: "If you pay homage to kings, do it like this."[12]

---

9 Rashīd al-Dīn Faḍlullāh Hamadānī, in Hillenbrand, *Turkish Myth*, 92–93.

10 al-Ṭurṭūshī, in Hillenbrand, *Turkish Myth*, 29.

11 Hillenbrand, *Turkish Myth*, 117.

12 al-Jawzī, in Hillenbrand, *Turkish Myth*, 41. A more humiliating version in Sibṭ Ibn al-Jawzī, in Hillenbrand, *Turkish Myth*, 71.

This episode would certainly have reminded the Muslim audience of an even more famous incident charged with religious fervour, that between Guy of Lusignan and Saladin after the Battle of Hattin (1187). Furthermore, the use of a human skull as a drinking cup in ritual use or as a trophy is most often associated with the nomadic cultures of the Eurasian Steppe, hence the example of Khan Krum of the First Bulgarian Empire who was said to have made a jewelled cup from the skull of the Byzantine emperor Nikephoros I after killing him in the Battle of Pliska (811). The accounts of Manzikert stress the imposing strength of the Seljuk sultan and highlight the magnitude of the ensuing Muslim victory by having the captured enemy sovereign become his cup bearer, before he pulls his hair, a symbol of manliness and strength, and puts his face on the ground in another *proskynesis/adoratio*. In the account of al-Jawzī, the sultan also "struck him [Romanos] three or four blows with his hand and he kicked him a similar number of times."[13]

Nonetheless, after the humiliation comes the clemency and forgiveness, hence the sources make it clear that Alp Arslan released his royal prisoner after almost a week in captivity, having treated him honourably by the standards of the time. Before Saladin's magnanimity towards the defeated crusader leaders after Hattin, the Muslim audience would have been reminded of another great sultan, Alp Arslan, who had demonstrated to the world how a Muslim ruler shows mercy and restraint to a humbled enemy king.[14] We read the imaginary dialogue between the two heroes in al-Jawzī's account.[15] First, the sultan asks why Diogenes dismissed his envoys who brought offers of peace, done so to remind the Muslim audience that it was the emperor who chose the path to war, arrogantly thinking he could inflict an easy victory over the sultan:

> He [Alp Arslan] said to him: "Did I not give permission to the envoys of the caliph that they should seek you out, draw up peace terms with you and respond therein to your request? Did I not now send a message to you and make an offer to you that I would withdraw from you, and you refused and so on. What thing caused you to infringe [the treaty]?" He [Romanos] said: "I collected [troops], o Sultan, and I had superior numbers and I had the upper hand, but the victory was yours. So, do what you want and stop rebuking me."

To show once again that the sultan was more magnanimous and generous than his captured enemy, Alp Arslan allegedly asked Diogenes what he would have done if their situation was reversed: "He [Alp Arslan] said: 'If I had fallen into your hands, what would you have done with me?' He [Romanos] said: 'Something vile.' He [Alp Arslan] said:

13 al-Jawzī, in Hillenbrand, *Turkish Myth*, 40.

14 Hillenbrand, *Turkish Myth*, 121.

15 al-Jawzī, in Hillenbrand, *Turkish Myth*, 40. Compare with: al-Ṭurṭūshī, in Hillenbrand, *Turkish Myth*, 29; al-Ḥusaynī, in Hillenbrand, *Turkish Myth*, 56–57; al-Athīr, in Hillenbrand, *Turkish Myth*, 66; Sibṭ ibn al-Jawzī, in Hillenbrand, *Turkish Myth*, 70–71; al-ʿAdīm, in Hillenbrand, *Turkish Myth*, 76–77.

'He has spoken truthfully, by God! If he had said otherwise, he would be lying. This is an intelligent, strong man. It is not fitting that he should be killed.'"

The sultan even asked the prisoner himsef what he should do with him, with the emperor offering three options, with the first two being death or being publicly paraded around the sultan's empire as a punishment. The third option would have probably caused laughs amongst a Muslim audience, as there is a touch of humour when Romanos suggests to Alp Arslan that he could also set him free, although acknowledging it is not at all likely. But, the sultan does just that:

> He [Alp Arslan] said [to Romanus]: "What do you think should be done with you now?" He [Romanus] said: "One of three things: the first is to kill me, the second is to parade me publicly in your country which I almost attacked and captured, and [as for] the third–there is no benefit in mentioning it, for you will not do it." He [Alp Arslan] said: "Mention it." He [Romanos] said: "Pardoning me, accepting money, ransoming me, attaching me to your service and sending me back to my kingdom as a mamluk of yours, as a deputy for you in the land of Byzantium." He [Alp Arslan] said: "I have decided in respect of you only on that which suits your hopeless position. After thinking about it, bring enough money for your release and I will set you free."

A final symbolic act of subservience to the sultan was Alp Arsland's return of Diogenes's "outer garment [imperial crimson robes] and his hat and he dressed him in them with his own hand, saying: 'I have treated you well and I am satisfied with your loyalty. I am sending you to your country and returning you to your kingdom.'"[16] Interestingly, in the *Saljuq-nama* of Zahir al-Din Nishpuri (d. ca. 1186–1187/582), Alp Arslan put a ring in both Romanos's ears, bearing in mind that such rings were an ancient sign of servitude in the Middle East.[17]

A Treaty Agreement

Clearly Alp Arslan saw his captive as a valuable diplomatic asset. Once the symbolic punishment was over, the Seljuk sultan demanded a ransom of ten million dinars that, according to Sibṭ ibn al-Jawzī who probably had access to the official report of these negotiations that was subsequently sent to the 'Abbasid Caliph's court, was impossibly high.[18] A sum of half a million plus three hundred sixty thousand in annual tribute was eventually agreed.[19] Furthermore, Romanos agreed to release all Muslim prisoners in Byzantine hands and not to interfere in the lands of Islam in the future. The emperor even agreed to hand over to the Seljuks the strategic cities of Antioch, Edessa, Hierapolis, and Manzikert, and he went as far as to promise that "if I arrive safely in my country, I

16 Sibṭ ibn al-Jawzī, in Hilligrand, *Turkish Myth*, 72.

17 Nishpuri, in Hilligrand, *Turkish Myth*, 36.

18 Nicolle, *Manzikert* 1071, 86.

19 Sibṭ ibn al-Jawzī, in Hilligrand, *Turkish Myth*, 71.

will send troops to them [the people in those cities], I will lay siege to them [the cities] and I will take them [the cities] from them [the Byzantine governors] and I will hand them over to you." And, perhaps in the very amicable environment of the banquet held in honour of his prisoner/guest, Alp Arslan and Romanos may have even agreed on a marriage alliance between their children, though this never happened.[20]

Political Turmoil and Civil Wars

The consensus among modern historians is that the real cause behind the disintegration of Byzantine power in Anatolia lay with the ambitions of the Norman renegade mercenary leader, Roussel of Bailleuil, who sought to profit from the political turmoil and instability that followed the captivity and subsequent deposition of Romanos Diogenes, and establish his own polity in northeastern Asia Minor. The key milestone for the future of central and eastern Anatolia were the two Byzantine campaigns to subdue the Norman leader, mounted in 1073 and 1074. These campaigns, combined with the later civil wars of 1077–1081, eventually transformed Byzantine politics in the region into a game of aligning Turkish interests from which the weak Byzantine governments would emerge as the great losers.

The first critical challenge the Byzantine Empire had to confront in the aftermath of Manzikert lay in the mad scramble for power in Constantinople. With the Doukas family in the capital using the Varangian Guard to proclaim Romanos Diogenes deposed, the new regime—headed by the *kaisar* Ioannes, the younger brother of Emperor Constantine X Doukas (r. 1059–1067), and his sons Andronikos, the "survivor" of Manzikert, and Constantine—tonsured the Empress-regent Eudokia (ca.1021–1096) who proclaimed her son, and Ioannes's nephew, Michael as sole emperor.[21] The Doukas's next step was to dispatch an army under Constantine Doukas against Romanos, when they learned that the latter had been released from Turkish captivity, was collecting taxes in the Armeniakon theme, and had reached Dokeia (modern Tokat, 100 km northwest of Sebasteia and 55 km southwest of Neokaisarea).[22] According to Attaleiates, the imperial units under Alyates had sought refuge there after Manzikert and are reported to have pledged their allegiance to the emperor. However, Romanos's defeat in battle forced him to retreat to Kappadokia, and then to Cilicia, where he spent the winter of 1071/72 under the protection of the *doux* of Antioch, Chatatourios.[23] Yet another defeat in the following spring (1072) compelled the former emperor to appeal for help to the Seljuks, while the gov-

20 Bryennios, *Ὕλη Ἱστορίας*, 95–97.

21 Attaleiates, *History*, 305–9. Bryennios is the only one who mentions the compliance of the Varangian Guard to the new regime: *Ὕλη Ἱστορίας*, 97–99; Zonaras, *Επιτομή*, 3:221. See also: Kaldellis, *Streams of Gold*, 249–50; Treadgold, *History*, 603.

22 Attaleiates, *History*, 301–3, 309; Skylitzes, *Continuation*, 129–31; Cheynet, "Mantzikert," 429; Haldon, *The Byzantine Wars*, 180.

23 Leveniotes, *Η πολιτική κατάρρευση του Βυζαντίου*, 603 (Table IZ.1.266). Chatatourios was replaced by the new Doukas regime with none other than the protoproedros Joseph Trachaneiotes, promoted from magistros shortly after Romanos's blinding.

ernment of the Doukas had already acquired the services of the Frankish mercenary leader Crispin and his followers. In the end, the lack of reinforcements and hope forced the embattled Romanos to surrender to the mercy of his enemies. However, no mercy was spared—the emperor was cruelly blinded on June, 29, 1072.

The callous treatment of Romanos by his enemies provided Alp Arslan and his emirs with a pretext to break the treaty with the empire, and the various Turkish groups that infiltrated Asia Minor after 1072 began to spend extended periods, or even settle permanently in several regions of central and eastern Anatolia, as they no longer had any fear of the military forces of the empire.[24] We should consider that it was probably around this time that two of the most well-known chieftains, identified as Artuk and Tutach, started to make inroads into Anatolia.[25] The available sources do not reveal exactly when or how the towns of Anatolia fell into Turkish hands, but Rashīd al-Dīn Faḍlullāh Hamadānī reports that Alp Arslan, after learning of Diogenes's fate, "gave orders to the commanders that they should venture deep into the lands of Byzantium, that each territory or city they succeeded in seizing and acquiring, each one should have it." More specifically, the family and followers of the emir Saltuk "took Theodosiopolis and its adjoining dependencies;" the Artukids "took Mardin, Amida, Manzikert, Melitene, Kharput" and the neighbouring regions; the Danişmendids "took Kaisarea, Zamandu [Zamanti River that crosses central Anatolia, southeast of Kaisarea], Sebasteia, Davalu [?], Dokeia, Neokaisarea and Amasea;" Çavuldur subdued Germanikeia [modern Maraş], Mengücek took Keltzinē [modern Erzincan], and Kamakha [mid-way between Keltzinē and Tephrike in the Euphrates valley].[26] The exact number of Turks who infiltrated into Anatolian regions after 1071 is, of course, unknown, although Mükrimin Halil Yinanç published a highly controversial estimate of (a well-rounded) one million Turks that penetrated into the interior of the peninsula during the first phase of Turkish settlement in Asia Minor; however, this number is hypothetical and greatly exaggerated.[27]

What came to add significantly to the socio-political unrest in Anatolia in the aftermath of Manzikert was Roussel of Bailleuil's attempt to establish a realm for himself in Asia Minor. Frankish warriors from Italy first appear as individual mercenaries in the service of the Byzantine State as early as 1047, and they are found in every major operational theatre in the Balkans and in Anatolia throughout the 1050s and 1060s, fighting—probably on horseback—alongside regiments of the elite Varangian Guard.[28] Roussel of Bailleuil was perhaps the most (in)famous of the Franks to have been employed

24 Skylitzes, *Continuation*, 139; Zonaras, *Επιτομή*, 3:227; Attaleiates, *History*, 333.

25 Bryennios, *Ύλη Ιστορίας*, 163–67 (for Artuk), 171–73 (for Tutach); Leveniotes, *Το στασιαστικό κίνημα*, 149–53 (for Artuk), 173–75 (for Tutach).

26 Rashīd al-Dīn Faḍlullāh Hamadānī, in Hillenbrand, *Turkish Myth*, 94–95.

27 Mükrimin Halil Yinanç, *Türkiye Tarihi Selçuklular Devri* (Ankara: Türk Tarih Kurumu, 2020), 159–61. View rejected in detail by: Vryonis Jr., *Η παρακμή του μεσαιωνικού ελληνισμού*, 510–11n269.

28 Theotokis, "Rus, Varangian and Frankish Mercenaries," 126–56; Alicia Simpson, "Three Sources of Military Unrest in Eleventh Century Asia Minor: The Norman Chieftains Hervé Frankopoulos, Robert Crispin and Roussel of Bailleuil," *Mesogeios/Méditerranée* 9–10 (2000): 181–207; Jonathan

by Constantinople. He was Roger Hauteville's principal lieutenant in Sicily, who won the day for the Normans at the Battle of Cerami in June 1063.[29] The next mention of Roussel comes in 1071 during Emperor Romanos's fatal campaign at Manzikert, where Roussel's contingent numbered around five hundred men. Then, he followed the "treacherous," "greedy" and "violent" nature that characterized all Frankish and Norman warriors, according to Anna Komnene, and he rebelled against his paymasters.

Roussel's rebellion is described in detail by our Greek sources.[30] The Frank sought an opportunity to break with the government in Constantinople when he was ordered to take part in a campaign led by Isaakios Komnenos (Alexios's elder brother and newly appointed *Domestikos of the East*) to "pacify" Anatolia in 1073. This expeditionary force included some four hundred Frankish horsemen under Roussel who, under the pretext of protecting one of his men who had been mistreated by Byzantine officers, he split with the main body of the imperial forces at Ikonion. Isaakios was eventually defeated and captured by the Turks close to Kaisarea, prompting some modern historians to speculate what would have happened if the Byzantine commander have had Roussel's elite heavy cavalry unit at his disposal.[31] My impression is that Isaakios would have achieved little in terms of strategic and battlefield manoeuvrability, because the Turkish horse-archers would have easily kept out of reach of the Frankish lances, as the crusaders experienced to their horror a quarter of a century later at Dorylaion/Doryleum.[32] What is certain, however, is that Isaakios's army was the last Christian army that would march across Asia Minor to Kaisarea—in the heart of the Anatolian Plateau—until the coming of the First Crusade.

For the imperial campaign of 1073, Bryennios extols the efforts of Alexios Komnenos, who had participated in the campaign alongside his elder brother Isaakios, to raise the morale of the soldiers and to organize the defence of the Byzantine camp against the Seljuks. He also reports of Alexios's visit to Ankyra to investigate the rumours that his brother was held hostage there—the latter mention of the "metropolis of the Galatians" (i.e., Ankyra) would be the last one before the capture of the city by the Seljuks, as it appears in the *Ὕλη Ιστορίας* to be under severe military pressure by the Turks.[33] Isaakios's defeat and capture close to Kaisarea, in 1073, was followed by Alexios's pursuit of his brother's captors in a northwesterly direction as far as Ankyra, old capital of Galatia and the Boukellarion theme, and the ambush—soon after Isaakios's release—of the

---

Shepard, "The Uses of the Franks in Eleventh-Century Byzantium," *Anglo-Norman Studies* 15 (1993): 275–305.

29 Georgios Theotokis, "The Norman Invasion of Sicily, 1061–1072: Numbers and Military Tactics," *War in History* 17 (2010): 381–402.

30 Attaleiates, *History*, 333–51; Skylitzes, *Continuation*, 139–45; Bryennios, *Ὕλη Ιστορίας*, 125–27, 149–57; Zonaras, Επιτομή, 3:227–29, 235.

31 Kaldellis, *Streams of Gold*, 258; Beihammer, *Byzantium and the Emergence of Muslim-Turkish Anatolia*, 207–9.

32 Georgios Theotokis, "The Square Fighting March of the Crusaders at the Battle of Ascalon (1099)," *Journal of Medieval Military History* 11 (2013): 57–72.

33 Bryennios, *Ὕλη Ιστορίας*, 127–31 and 137; Leveniotes, *Το στασιαστικό κίνημα*, 110–15.

two Komnenoi brothers by a band of two hundred Turks outside Nikomedia (less than 100 km east of Constantinople). This clearly shows that the invaders had opened the way to Bithynia and the Marmara Sea through the southwestern areas of the Anatolian plateau from Kappadokia to Lykaonia, Phrygia, and the Upper Meander and Sangarios River Valleys.

In the spring of 1074, an expedition was dispatched against Roussel under the *kaisar* Ioannes, who had also brought with him Varangian and Frankish units. But they were defeated at the famous Battle of the Zombou Bridge over the Sangarios River, with the imperial Franks deserting to Roussel before the battle.[34] Leveniotes makes the pertinent point here that "the extraordinary victory of the small in numbers army at the Battle of the Zombou Bridge humiliated the imperial army, demolishing the last vestiges of its prestige after Manzikert." Picking up from his triumph, the Frankish captain then attempted something unimaginable: to take his troops—by now numbering some three thousand in all—and march to Constantinople in the hope that the people of the capital would open the gates to him. Sometime at the end of 1074, Roussel even tried to create his own "puppet-emperor" by proclaiming the captive *kaisar* Ioannes as the legitimate ruler of the empire, in an attempt to provide political cover to his aggression;[35] this tactic would be adopted by Robert 'Guiscard" Hauteville seven years later. Roussel's illusions would quickly be dashed.

The growing power of Roussel's forces compelled the Doukas regime to seek a strong counterweight, which at that time only the Turks could provide.[36] Hence, the chief minister of the imperial government took the fateful decision to hire a Turkish marauder in Bithynia, a certain Artuk we saw earlier [usually identified with Abu Sa'id Taj ad-Dawla Tutush, the younger brother of the Seljuk Sultan Malik Shah I (r. 1072–1092)].[37] The Turks defeated Roussel's forces at the Battle of Sophon (1074/75) by applying their typical steppe tactics of feigned retreat while picking off the Franks with their arrows from a safe distance;[38] an outcome that reinforces my last point about Isaakios Komnenos's defeat by the Turks the previous year. Henceforth, the Byzantines would hire Franks to fight Turks and Turks to fight Franks.

As Roussel resumed his operations in the Armeniakon theme, collecting taxes and extorting protection from the locals,[39] the Doukas regime now gave the task of capturing Roussel to Nikephoros Paleologos. In 1075, Nikephoros led a considerable force of Byzantine and Georgian troops against the Franks, but his men deserted him after he

---

34 Leveniotes, *Το στασιαστικό κίνημα*, 121–29.

35 Leveniotes, *Το στασιαστικό κίνημα*, 140–42.

36 Leveniotes, *Το στασιαστικό κίνημα*, 144–50.

37 Leveniotes, *Το στασιαστικό κίνημα*, 149–50.

38 Attaleiates, History, 345–49; Skylitzes, *Continuation*, 141–43; Bryennios, *Ὕλη Ιστορίας*, 151–57; Zonaras, *Επιτομή*, 3:227; Leveniotes, *Το στασιαστικό κίνημα*, 151–53.

39 "For the cities which obeyed him [Roussel] before, paying tribute more out of fear than out of prudence, when they began to doubt his position, all began to give him nothing," Bryennios, *Ὕλη Ιστορίας*, 171; Leveniotes, *Το στασιαστικό κίνημα*, 157–67.

failed to pay them on time.[40] Then, probably in 1076, it was Alexios Komnenos's time to appear in the political scene at the young age of eighteen; Artuk had treacherously arrested Roussel and delivered him to Alexios at Amaseia, where the latter simply took him as prisoner back to the capital.[41]

The critical phase for the permanent establishment of Turkoman bands in western Asia Minor roughly coincides with the years between the revolt of the *dux* of Anatolikon Nikephoros Botaneiates, in June 1077, and the revolt of Alexios Komnenos, in April 1081. Undoubtedly, this was a period of massive Turkish involvement in Byzantine internal conflicts and power struggles, during which the Turks sought the opportunity to make money and gain privileges and land. From the outset of Botaneiates's revolt, it is clear that both Michael VII and Nikephoros were keen to employ the Turkish warriors roaming in northwestern Asia Minor, and Suleiman b. Qutlumush and his men played a crucial role in these negotiations. According to Attaleiates, it was God's providence that turned Suleiman to agree to an alliance with the rebel general, and it is likely that Botaneiates promised Suleiman a strategic town as a seat of his ever-extending power in the region: Nicaea.[42]

Suleiman's Turks were also involved in the seditious movement of the *dux* of Dyrrachium Nikephoros Bryennios ("the Elder"), the 'survivor" from Manzikert, who—around the same time as Botaneiates—had risen against Michael VII having the support of the Byzantine army's regiments in the Balkans. Bryennios continued his revolt even after Botaneiates was crowned emperor in March 1078, and after failed negotiations, the young Alexios Komnenos was dispatched against him. Interestingly, Alexios sought a significant force of two thousand horsearchers from "the Turkish chieftains residing in Nicaea" (i.e Suleiman and, his brother, Mansour),[43] to take part in an expedition that would culminate in the Battle of Kalavrye, by the Halmyros River in Thrace, where the rebel army was broken and Bryennios was captured.[44]

Two more revolts followed Kalavrye, and both of them were crushed by the up-and-coming *Domestikos of the Scholae* of the West, Alexios Komnenos. Another veteran of Manzikert, Nikephoros Basilakes brought up the armies of Dyrrachium and Bulgaria and some Franks from Italy to Thessaloniki, only to be defeated by a cunning trick by Alexios. Two years later, in the autumn of 1080, Nikephoros Melissinos, another general who had remained loyal to Michael VII and had been exiled by Botaneiates to the Aegean Island of Kos, decided to raise an army of Turks from western Asia Minor and march

40 This campaign is only reported by Bryennios, *Ὕλη Ιστορίας*, 167–69. Kaldellis, *Streams of Gold*, 259; Leveniotes, *Το στασιαστικό κίνημα*, 167–68.

41 *Alexiad*, bk. 1, 1–3; Bryennios, *Ὕλη Ιστορίας*, 173–75; Leveniotes, *Το στασιαστικό κίνημα*, 169–89.

42 Attaleiates, *History*, 433–41; Zonaras, *Επιτομή*, 3:235–39; Beihammer, *Byzantium and the Emergence of Muslim-Turkish Anatolia*, 218–9; Kaldellis, *Streams of Gold*, 265; Cahen, *Pre-Ottoman Turkey*, 75–76.

43 Bryennios, *Ὕλη Ιστορίας*, 257–61.

44 Norman Tobias, "The Tactics and Strategy of Alexius Comnenus at Calavrytae, 1078," *Byzantine Studies* 6 (1979): 193–211.

to Constantinople. Melissinos was married to Alexios's elder sister Eudokia, hence the former's refusal to face him in a pitched battle when ordered to by Botaneiates. Between the autumn of 1080 and the spring of 1081, Melissinos seized control of what remained of Byzantine Asia Minor with Turkish aid and proclaimed himself emperor. However, following the successful seizure of the throne by his brother-in-law Alexios Komnenos on 1 April 1081, he submitted to him, accepting the title of *kaisar* and the office of the governor of the second largest city of the empire, Thessaloniki.

Some Conclusions

Although the period between 1073 and 1076 marks the definitive loss of eastern and central Asia Minor to the Turks, who successfully took advantage of a power vacuum and offered their services to all parties involved, it was ironically Roussel's Franks that were perceived to be a more immediate threat for the Doukas regime, especially in view of Roussel's proclamation of *kaisar* Ioannes as "emperor."[45] The critical phase for the permanent establishment of the Turks in western Asia Minor roughly coincides with the years between the revolt of the *doux* of Anatolikon Nikephoros Botaneiates in June 1077, and the revolt of Alexios Komnenos in April 1081, key conflicts for which it is clear that every participant was keen to employ the Turkish warriors roaming in northwestern Asia Minor.[46]

---

45 Beihammer, *Byzantium and the Emergence of Muslim-Turkish Anatolia*, 207–15.

46 Kaldellis, *Streams of Gold*, 265; Beihammer, *Byzantium and the Emergence of Muslim-Turkish Anatolia*, 218–9; Cahen, *Pre-Ottoman Turkey*, 72–76.

CONCLUSIONS

The Oriental conquerors were not inclined to hurry; the whole process would take them the best part of four centuries. But there can be no question that the advance that ended on Tuesday, 29 May 1453, when Sultan Mehmet II touched his turban to the floor of St Sophia in prayer and thanksgiving, had its origin on the distant field of Manzikert, 382 years before.[1]

A Military Disaster?

In a ground-breaking article published in 1980, Cheynet pondered whether Manzikert was "un désastre militaire."[2] Since then, the consensus among historians of the period has been that Manzikert was *not* a military disaster and that the imperial army seems to have escaped relatively unscathed from the day's fighting on August 26, 1071, with the political and socio-economic consequences of the defeat far outweighing the losses the empire sustained on the battlefield.[3] The rearguard and reserve units of "the tagma of the *Hetairiae* and that of the nobles" under Andronikos Doukas certainly escaped back to the capital without suffering any casualties, whilst Bryennius's left wing, which included the five western tagmata also escaped with relatively few losses; these units are reported defending the Balkans against the Patzinaks in the following years. Concerning the units of the right wing under Alyates, and especially the Armenian and Kappadokian forces, the sources tell us that a significant number of them managed to make an orderly escape to Trapezounta, Theodosiopolis, and Dokeia.[4] According to Attaleiates, the units under Alyates reportedly pledged their allegiance to the emperor after his release from Turkish captivity.

Furthermore, what is notable is that the prominent men who were reported to have died in the battle were few and relatively unknown figures (Leon, who was *epi ton deeseon* and the *magistros* and *protoasecretis* Eustratios Choirosphaktes), while even

1 John Julius Norwich, *Byzantium: The Apogee* (London: Penguin, 1993), 3.

2 Jean Claude Cheynet, "Mantzikert: Un désastre militaire?," *Byzantion* 50 (1980): 410–38.

3 Kaldellis, *Streams of Gold*, 248–49, 271–72; Peter Frankopan, *The First Crusade: the Call from the East* (Cambridge, MA: Harvard University Press, 2012), 31–32; Hillenbrand, *Turkish Myth*, 15–17; Julian Chrysostomides, "The Byzantine Empire from the Eleventh to the Fifteenth Century," in *The Cambridge History of Turkey; V. 1: Byzantium to Turkey, 1071–1453*, ed. Kate Fleet (Cambridge: Cambridge University Press, 2009), 10–12; Nicolle, Manzikert 1071, 83–86; Beihammer, *Byzantium and the Emergence of Muslim-Turkish Anatolia*, 161; Leveniotes, *Η πολιτική κατάρρευση του Βυζαντίου*, 180–81; Angold, "Belle Époque or Crisis?," 609; Angold, *The Byzantine Empire*, 46; Haldon, *Warfare, State and Society*, 225–26; Treadgold, *History of the Byzantine State and Society*, 603–4. Vryonis Jr. does not refer to the military losses sustained during the battle, and rather emphasizes the "internal developments within the empire" that followed: *Η παρακμή του μεσαιωνικού ελληνισμού*, 92–93.

4 Attaleiates, *History*, 301–3, 309; Skylitzes, *Continuation*, 129–31; Cheynet, "Mantzikert," 429; Haldon, *The Byzantine Wars*, 180.

fewer were captured alive (the *protovestes* Basileios Maleses and Nikephoros Basilakes, the latter having been taken prisoner the day before the main battle).[5] If we add the elite units of Trachaneiotes and Roussel, which retreated to Melitene before the battle, a significant number of perhaps even twenty thousand men, it seems clear that the actual losses incurred during the battle were limited to the Diogenes's immediate retinue, the Armenian infantry, and the palatine and other elite units close to him, around five to ten percent of the campaigning army.[6]

To compare the aftermath of the Battle of Dyrrachium in October 1081 was far more devastating for the imperial army led by Alexios Komnenos than the outcome of the Battle of Manzikert was for the army under Romanos IV, yet the political consequences for the empire were very different. The losses for the Byzantine forces at Dyrrachium were certainly heavier, and it is possible that as much as a quarter of the total imperial forces engaged were killed or wounded. On top of the Varangian regiment that was annihilated that day, the Byzantine nobility (allied to the Komnenoi) also suffered a severe blow, with "several fine soldiers killed," such as the *porphyrogennitos* Constantine Doukas, Nikephoros Paleologos, general Aspietes, and Nikephoros Synadenos.[7]

However, following the subsequent siege and capitulation of the city of Dyrrachium to Robert Guiscard's forces on February 21, 1082, the harsh winters in the mountainous regions of Illyria and western Macedonia prevented the Norman duke from taking advantage of his battlefield success in the previous autumn. Meanwhile, having escaped the fate that befell several of his officers and men, Alexios Komnenos had already established Thessaloniki as a rallying point for the remnants of his troops returning from Illyria. Even though the imperial treasury was empty and could not afford the hiring of extra mercenary troops to defend against the Norman advance, the emperor resorted to a desperate and highly unpopular (although he had the support of the residential synod and the patriarch Eustathios Garidas (1081–1084) measure of confiscating precious ecclesiastical objects from various churches in the capital to fill the coffers.[8] The war was not over for the young emperor, it had merely begun!

A Decisive Battle

This was not the case for Romanos Diogenes! The aftermath of the Battle of Manzikert "should only have confirmed Turkish domination of the Armenian highlands, not that almost the whole of Anatolia would be overrun by the Turks within ten years,"[9] a pertinent point that helps to confirm my argument that Manzikert was indeed a decisive battle. At the end of the reign of Romanos's predecessor, Constantine X Doukas, the imperial armies appeared to be underfunded, demoralized, untrained, and disorganized, while

5 Attaleiates, *History*, 303; Attaleiates, *History*, 301–3.

6 Cheynet, "Mantzikert," 431; Haldon, *The Byzantine Wars*, 180.

7 *Alexiad*, bk. 4, 6; Theotokis, *Norman Campaigns*, 158, 165.

8 Angeliki Panagopoulou, "Nomos and Canon in Byzantium: The Case of the Confiscation of the Holy Vessels during the Reign of Alexios I Comnenus," *Byzantina Symmeikta* 31 (2021): 401–30.

9 Angold, "Belle Époque or Crisis?," 609.

the cities of eastern and central Asia Minor seemed to have been under the constant threat of being raided, sacked, and destroyed by the Turks. However, in the decades after the death of Basil II in 1025, the Byzantines did not all of a sudden become incompetent in the field of battle, but, as Vryonis Jr. notes, it was just that Byzantine defeats were more numerous and more *decisive* than their victories.

The vigorous campaigns of Diogenes show that there was the potential for the situation to be reversed and that numbers, organization, discipline, and strategic sense were still there. There is little doubt among historians today that Romanos IV was an able and competent commander, taking care to defend his camps, and moving his troops by logistical needs; that is why he was the preferred choice to marry Eudokia and become an emperor in the first place. Diogenes's campaign to "plug the Syrian gap" in 1068 can be viewed as a strategic success, especially if we consider the imperial appointment of a new imperial governor of Hierapolis who controlled the road network between Edessa and Antioch. Nonetheless, what we also see is the strategic initiative passing firmly into the hands of the Seljuks, with Diogenes and the imperial armies now reacting to and struggling to catch up with the Turkish marauders. The perceived Byzantine weakness in the "central" border region (i.e., Melitene, Sebasteia, and Kappadokia) gave rise to a significant increase of hostile activities in the Upper Euphrates and the Anti-Taurus regions for the campaigning season of 1069, yet the emperor seemed unable to inflict a decisive defeat on a Seljuk army.

The sack of Ikonion in 1069 was added to that of Amorion the previous year, and the Byzantine defences of the central Anatolian plateau admittedly seemed more porous every year. It is with this background in mind that we should consider the strategic decision behind the dispatch of Manuel Komnenos to set up his base of operations for the 1070 campaigning season in Kaisarea, on the eastern edge of the Anatolian Plateau, instead of Theodosiopolis or Antioch. This perceived shift in the strategic priorities of the Byzantine government suggests that the emperor had dropped his attempts to seal off the "central" border region of Taron, Vaasparakania, and Melitene. However, while Alp Arslan was pursuing his strategic goal outside Aleppo in the winter of 1070/71, Diogenes was launching his own campaign that targeted the strategic fortress-cities on the northern side of Lake Van, turning his attention to the "Armenian Gap" of the Arsanias River Valley.

The Byzantine campaign of 1071 that culminated at the Battle of Manzikert in August proves that the Byzantine side could reclaim the strategic initiative, and Romanos's thrust to Vaasparakania seemed to have caught Alp Arslan by surprise. Even if the sultan had adopted a defensive strategy and kept the bulk of his forces in Iran or Upper Mesopotamia, and there was no pitched battle for that campaigning season, then the Byzantines would have been able to 'plug the Armenian gap' unopposed, thus rendering the Arsanias valley that led to Melitene and Kaisarea too dangerous for the Seljuks. Nonetheless, there was a pitched battle between the two armies that provided history with a clear winner, Alp Arslan.

We already saw that Manzikert was not a disaster for the empire because of the huge loss of human life. But what about its impact on the region and beyond? To reiterate what I wrote in the introduction about what makes a battle decisive, the short and simple answer is its geopolitical impact, adding to that effect that a decisive battle has long-term socio-political implications that fundamentally change the balance of power on more than just the local level. But why should a defeat at the extreme limit of Byzantium's eastern frontiers have opened up Asia Minor to Turkish settlement? The answer is because of the element of chaos that Manzikert introduced in the equation of Byzantine politics and the Turkish penetration of Anatolia.

In the introduction to this study, I examined a critical reason that can explain the rarity of battles in the history of warfare–the presence of a sovereign in the field of battle. It was the dramatic technological advances in communications and telegraphy during the Second Industrial Revolution that allowed generals and chiefs of general staff during the First World War to issue orders to their units from their HQ several kilometres behind the front line, although these lines were also fragile and inevitably became the target of artillery attacks by the attackers.[10] Just half a century before at Sedan, on September 2, 1870, the French emperor Napoleon III was taken prisoner by the Prussian general Helmuth von Moltke and, although French casualties were minimal (130,000-strong French *Army of Châlons* sustained only 3,220 killed, with 104,000 men and 558 guns passing into captivity), the geopolitical impact of the Battle of Sedan (September 1–2, 1870) was *decisive*. Not only did the capture of the French emperor leave the Prussians without an opposing government willing to make a quick peace, but on the same day of Napoleon's capture, a group of protesters stormed the imperial palace and the French Second Empire collapsed in a bloodless revolution.[11]

To return to the eleventh century, I believe it is clear by now that in a period of history when a king or a sultan were at the forefront of fighting, the death or capture of a leader on the battlefield could have a *decisive* impact in the course of history. That is one of the main reasons why battles were rare, along with the element of chance/friction that always "loomed over" every battlefield. At Hastings, the "accidental arrow" through King Harold's eye was undoubtedly a development of tremendous importance for the future of the Anglo-Saxon kingdom that allowed William of Normandy to claim the entire kingdom as his own by right of conquest. At Manzikert, however, it was not just the defeat of the imperial army on the battlefield, but the capture of Emperor Romanos and, thus, the loss of legitimacy and political vacuum that ensued, that enabled Diogenes's enemies to move against him and set off a period of political instability.

After his defeat at the Battle of Dyrrachium, Alexios Komnenos went into a frenzy of diplomatic and military activity,[12] establishing Thessaloniki as a rallying point for the remnants of his troops returning from Illyria, confiscating sacred vessels to mint money in order to hire Turkish mercenaries from the Sultanate of Rum, setting in motion the mechanisms of Byzantine diplomacy by sending an embassy to the German emperor Henry IV in early April 1082, promising lavish gifts and a royal marriage, while an Apu-

10 "Fighting Talk: First World War Telecommunications," *The National Archives*, https://www.nationalarchives.gov.uk/first-world-war/telecommunications-in-war/, accessed May 30, 2023.

11 Michael Howard, *The Franco-Prussian War: The German Invasion of France, 1870–1871* (London: Routledge, 1990), 175–77.

12 Theotokis, *The Norman Campaigns*, 167–68.

lian revolt was stirred up with Byzantine money against Robert Guiscard. The point I wish to make here is that Alexios's power was not (seriously) threatened at any point during his wars against the Normans,[13] despite the fact he sustained two further defeats in Epirus in 1082. For the first years of his new regime, Alexios was astute enough to use members of his family to act as a "protective ring" around himself in key posts of power, with the placement of the Komnenoi matriarch Anna Dalassene at the head of the government during Alexios's absence receiving extensive commenting by Anna Komnene and Nikephoros Bryennios.[14] To compare, Diogenes's "marriage of necessity" to empress-regent Eudokia Makrembolitissa did not earn the general any friends—on the contrary, the Doukas clan was eager to destroy him after his first setback. But what would have happened if Diogenes was not captured at Manzikert, and was able to return to the capital in an orderly fashion at the head of his troops? Maybe he would have been able to launch another campaign in the spring of 1072, accepting the loss of Vaasparakania, but perhaps marching another army to Kaisarea or Sebasteia in an effort to establish some sort of a defensive line to shield central Anatolia. We will never know.

Manzikert introduced an element of *chaos* in (Anatolian) history, with "small inputs creating very large perturbations;" it ushered in the civil war of 1071–1072, the three-year war with Roussel (1073–1076), and the multiple civil wars of 1077–1081, while it also indicated to the Seljuks that Asia Minor was "up for grabs," to the point that large chunks of Asia Minor soon became subject to new lords as we saw in the previous chapter: the Danishmends in Sebasteia (Sivas), Kaisarea (Kayseri), and Amaseia (Amasea); the Mengüceks in Keltzinē, Kamakha, and Tephrike on the upper Euphrates; the Saltukids in Theodosiopolis (Erzurum); and perhaps the most dangerous of all was the Sultanate of Rum, under Suleiman b. Qutalmish I (r. 1077/8–1086). The latter had been established shortly before the accession of Alexios Komnenos, previously being "allies" and a vassal state to Nikephoros Botaneiates,[15] and it occupied most parts of Bithynia,

13 The first serious challenge to Alexios's power was, I think, mounted by Nikephoros Diogenes, the son of Romanos IV, in the summer of 1094, while the emperor embarked on his campaign in the Balkans. Frankopan, *The First Crusade*, 80–84.

14 Peter Frankopan, "Kinship and the Distribution of Power in Komnenian Byzantium," *English Historical Review* 122 (2007): 1–34 (particularly 7n25); Nicolas A. Oikonomides, "L'évolution de l'organisation administrative de l'empire byzantin au XIe siècle," *Travaux et mémoires* 6 (1976): 125–52. More recently, Frankopan challenged the long-held views that the new emperor began his reign by purging the aristocracy and promoting his own family, although it is beyond doubt that many of Alexios's relatives were indeed promoted to critical offices of the empire. What Frankopan is also right to emphasize is the matter of split loyalties between the Doukas and Komnenoi clans in the immediate aftermath of Alexios's coup d'état: "Re-Interpreting the Role of the Family in Comnenian Byzantium: Where Blood is not Thicker than Water," in *Byzantium in the Eleventh Century: Being in between: Papers from the 45th Spring Symposium of Byzantine Studies, Exeter College, Oxford, 24–26 March 2012*, ed. Marc D. Lauxtermann and Mark Whittow (Abingdon: Routledge, 2017), 181–96.

15 On the critical role of the "Qutlumush Turks" in ousting Michael VII from the throne in favour of Nikephoros Botaneiates: Beihammer, *Byzantium and the Emergence of Muslim-Turkish Anatolia*, 215–24.

large parts of Phrygia and Galatia, and the Aegean coasts as far south as Phocaea, and its capital was Nicaea, situated just 40 km from Constantinople.

In the last years of Alexios's reign and during the reign of his son, John II (r. 1118–1143), substantial regions of western Asia Minor were restored to the empire; in 1117, Alexios pushed his army deep into the Turkish-dominated Anatolian Plateau, where he defeated the Seljuk sultan at the Battle of Philomelion (modern: Akşehir), while John II's expansion in the Maeander valley in western Anatolia (1119: Laodicea, 1120: Sozopolis) aimed to expel the Turks from Caria and unite western Byzantine Anatolia with Byzantine Attaleia. John's son Manuel I (r. 1143–1180) began gradually to extend Byzantine power into central Anatolia, defeating the Turks at Akroënos in 1146, before capturing and destroying the fortified town of Philomelion, and ravaging the area around the city of Ikonion. By the end of Manuel's reign, the Byzantine provinces in Anatolia stretched from Chaldea and Trabzon on the Pontic coast westwards in an arch through the districts of Paphlagonia, Optimaton, Nikomedia, Opsikion, Neokastra, Thrakesion, Mylasa/Melanoudion, Kibyrrhaiot, and Cilicia.

Manuel may have even considered the recovery of Kappadokia in the last years of his reign, an achievement which would have made possible the encirclement of the Seljuk Sultanate of Rum based at Ikonion as well as the Danishmendid emir to their north.[16] However, a strategically premature and tactically misjudged attempt to gain the upper hand over the Seljuks of Ikonion in central Anatolia resulted in the disastrous campaign that culminated at the Battle of Myriokephalon on September 17, 1176.[17] And although there are obvious, if superficial, parallels between Manzikert and Myriokephalon,[18] the latter was the last imperial inroad into the central Anatolian Plateau, because the later campaigns of the Normans of Sicily in the Balkans and the arrival of the Fourth Crusade ended forever any hope of a Byzantine recovery of the central and eastern regions of Asia Minor that had been lost a century before in the aftermath of Manzikert.

The Reception of the Battle of Manzikert in Modern Greek Historiography

The development of Byzantine studies in Greece essentially began in the mid-nineteenth century with the Greek historians Spyridon Zambelios and Konstantinos Paparrigopoulos and was due to two parallel phenomena: the renewed interest in medieval studies in Europe with *Romanticism*, and the search for national identity and the historical unity of *Hellenism* in the newly liberated Greece.[19] Konstantinos Paparrigopoulos (1815–1891) was a Greek historian who is considered by modern historians as the "father" of modern

16 Haldon, *Warfare, State and Society*, 97.

17 Haldon, *Byzantine Wars*, 192.

18 Haldon, *Warfare, State and Society*, 96–97; John Birkenmeier, *The Development of the Komnenian Army: 1081–1180* (Leiden: Brill, 2002), 126–35; Paul Magdalino, *The Empire of Manuel I Komnenos, 1143–1180* (New York: Cambridge University Press, 1993), 98–99; Angold, *The Byzantine Empire*, 224–25.

19 Dionysios A. Zakythinos, "Le monde de Byzance dans la pensée historique de l'Europe a

Greek historiography.[20] Together with Zambelios, they demolished with scientific arguments the theory of racial discontinuity of the modern Greeks,[21] formulated in 1830 by the Austrian historian Jacob Philipp Fallmerayer.[22] Paparrigopoulos is the founder of the concept of the historical continuity of Greece from antiquity to the present day since he established in his teaching at the University of Athens the tripartite division of Greek history (ancient, medieval, and modern) and sought to refute the dominant views of the time that the Byzantine Empire was a period of decline of *Hellenism*. A major reason why Paparrigopoulos considered Byzantium important was because it achieved the political unity of the Greeks, which was absent in ancient Greece, while his views on national unity acknowledge the influence of the *Megali Idea*,[23] something for which his contemporary Zambelios states that the "true" borders of neo-Hellenism were drawn many centuries before by Nikephoros Phokas and Ioannes Tzimiskes.[24]

In 1860, Paparrigopoulos began publishing the *Ιστορία του Ελληνικού Έθνους* [History of the Hellenic Nation], a work that established him in the scientific field. This work was divided into three volumes of fifteen books and its publication ended in 1876. For him, Byzantine history is the history of the Greek people—or rather of the collective historical agent, the Greek nation during the medieval millennium. In narrating the chronicle of the Byzantine millennium, the author's interest remains focused on ethnological issues, because his deeper concern is to depict the survival and continuous existence of that exceptional historical actor, the "Hellenic ethnos."[25] Paparrigopoulos's work, however, is not merely a lament for Byzantium, it is primarily a Greek epic, hence his depiction of the "epic" battle that shaped Byzantium and medieval Hellenism, the Battle of Manzikert. For Paparrigopoulos, this is a titanic struggle for the Hellenic empire to stay alive amidst the "flood" of foreign invasions from the East; he carefully desisted from using terms such as disaster, decline, or tragedy, although his depiction of

partir du XVIIe siècle: 1. Du Romantisme au Nationalisme," in *Byzance. Etat, société, économie*, ed. Dionysios A. Zakythinos and Hélène Ahrweiler (London: Variorum, 1973), 41–87.

20 K. Th. Dimaras, *Κωνσταντίνος Παπαρρηγόπουλος* (Athens: MIET, 1986)

21 The output of his interest in medieval history was his great work: *Αί Βυζαντιναί Μελέται. Περί πηγών Νεοελληνικής Εθνότητος από Η΄ άχρι Ι΄ εκατονταετηρίδος μ.Χ* (1857). See also: *Γιάννης Κουμπουρλής, Οι ιστοριογραφικές οφειλές των Σπ. Ζαμπέλιου και Κ. Παπαρρηγόπουλου. Η συμβολή Ελλήνων και ξένων στη διαμόρφωση του τρίσημου σχήματος του ελληνικού ιστορισμού (1782–1846)* (Athens: National Hellenic Research Foundation, 2012); Κουμπουρλής, "Οι οφειλές του Σπυρίδωνος Ζαμπελίου στη γαλλική ρομαντική ιστορική σχολή," in *Η΄ Διεθνές Πανιόνιο Συνέδριο*. Πρακτικά, *Χώρα Κυθήρων*, 21–25 Μαΐου, *Τόμος IVα: Επτανησιακός πολιτισμός (Μέρος Α΄)* (Athens: Εταιρεία Κυθηραϊκών Μελετών, 2009), 431–56.

22 Γιώργος Βελούδης, *O Jakob Philipp Fallmerayer και η γένεση του ελληνικού ιστορισμού* (Athens: E.M.N.E Μνημών, 1982)

23 Paschalis M. Kitromilides, "On the Intellectual Content of Greek Nationalism: Paparrigopoulos, Byzantium and the Great Idea," in *Byzantium and the Modern Greek Identity*, ed. David Ricks and Paul Magdalino (London: Routledge, 1998), 25–33.

24 George Huxley, "Aspects of Modern Greek Historiography of Byzantium," in *Byzantium and the Modern Greek Identity*, 18.

25 Kitromilides, "On the Intellectual Content of Greek Nationalism," 29–30.

the two "heroes," Alp Arslan and Diogenes has one aim in mind, which is portrayed in a sub-title of the main chapter about the Turkish invasions of Anatolia: "Romanos's Heroism, Captivity and Redemption." For Paparrigopoulos, Diogenes may have lost the battle, but he is the true hero of the "epic" story for the defence of medieval Anatolia, almost a "martyr-like" figure.[26]

The historian Pavlos Karolides (1849–1930) was a student of Paparrigopoulos. Of Kappadokian origin, Karolides succeeded his teacher in the chair of Greek history at the University of Athens in 1893. As he was an Ottoman citizen, he was elected to the Ottoman Parliament in 1908 as a representative of the Greeks of Sebasteia. He published his *Εγχειρίδιον βυζαντινής ιστορίας μετά των κυριοτάτων κεφαλαίων της μεσαιωνικής ιστορίας: προς χρήσιν των φοιτητών της Φιλολογίας* [Booklet of Byzantine History with the Main Chapters of Medieval History: For Use by the Students of Philology] (Athens: Tzakas, 1906) and in it, he also refers to Diogenes as a "heroic defender" of the "Greek lands in Asia Minor."[27] In another short essay focusing on emperor Romanos that he published in the same year, titled *Ο αυτοκράτωρ Διογένης ο Ρωμάνος* [Emperor Romanos Diogenes] (Athens: Σύλλογος προς Διάδοσιν Ωφελίμων Βιβλίων, 1906)], Karolides portrays Romanos as "the last upholder of Hellenism on the eastern frontiers of the empire,"[28] the last emperor to give spirit and power to the ill-disciplined crowd then called the Hellenic army, while after his deplorable end came the collapse of Hellenism in Asia Minor.[29]

Konstantinos Amantos (1874–1960) was the first professor to teach in the newly established (1924) chair of Byzantine History at the University of Athens. Apart from his major compositions covering the Byzantine and post-Byzantine periods, he wrote numerous works in the following categories: linguo-geographical concerning place names and etymological issues; historical (on the period of the Turkish occupation, the history of Chios, the history of education and scholars; the history of the Byzantine period); and ethnological, dealing with the history of the peoples neighbouring Greece, the Slavs, Albanians, and Turks, and their relations with each other. He published, among others, *Ο Ελληνισμός της Μικράς Ασίας κατά τον Μεσαίωνα* [The Hellenism of Asia Minor in the Middle Ages] (Athens: Sideris, 1919)], in a critical period in the Greek–Turkish relations and while the Greek army was campaigning in central Anatolia against Mustafa Kemal's troops, and the *Σχέσεις Ελλήνων και Τούρκων από του ενδεκάτου αιώνος μέχρι του 1821* [Greek–Turkish Relations between the Eleventh Century and 1821] (Athens: Ο.Ε.Σ.Β., 1955). In the latter work, Amantos portrays the army of Diogenes as predominantly Greek, with the (predominantly) Greek population of Asia Minor "longing for its relief from the Turkish depredations and raids," to add further down that "the defeat (of the Byzantine army) at Manzikert opened the way not just for the looting, but also

26 Konstantinos Paparrigopoulos, *Ιστορία του ελληνικού έθνους: Από αρχαιοτάτων χρόνων μέχρι σήμερα* (Athens: 4π, 2010), 14:223–30.

27 Karolides, *Εγχειρίδιον Βυζαντινής ιστορίας*, 228.

28 Karolides, *Ο αυτοκράτωρ Διογένης ο Ρωμάνος*, 74–75.

29 Huxley, "Aspects of Modern Greek Historiography of Byzantium," 20.

for the conquest of Asia Minor. The outcome of the defeat (at Manzikert)] can be felt to this day (1955). Since then and for over a thousand years, Hellenism has been fighting against the victors of Diogenes and it was its destiny, after so many struggles, to be uprooted from Asia Minor."[30]

A conventional view in the historiography of modern Greece tends to connect the origins of the Megali Idea and Greek nationalism with attitudes and aspirations dating from the late Byzantine period, specifically from the period of decline after the battle of Manzikert in 1071, the capture of Constantinople by the Crusaders in 1204, and the final Fall of Constantinople in 1453, which witnessed the rise of millenarian hopes for the recovery of the City by the Orthodox heirs of the Eastern Roman Empire.[31] This view has been propounded by distinguished Byzantinists such as Dionysios Zakythinos, Speros Vryonis Jr., Ioannes Karagiannopoulos, and Hélène Ahrweiler.

Dionysios Zakythinos (1905–1993) was a professor of the history of the Byzantine State at the University of Athens from 1939 to 1970 and was also the first director of the Centre for Byzantine Research in Athens (1960–1975).[32] In his *Βυζαντινή Ιστορια 324–1071* [Byzantine History, 324–1071] (Athens: Dodone, 1972), there is a clear distinction between the period of the Byzantine épopée of the Macedonian dynasty that ended with the death of Basil II, and the "decline" of the empire between 1028 and 1071 and the Battle of Manzikert. Zakythinos argues that "the defeat at Manzikert was one of the greatest in Byzantine history, and one of the most important events in the history of Europe," while he also adds that "Manzikert was a devastating defeat for the Byzantine army," as a consequence of which the empire "abandoned intact the Greek Asia Minor." For the historian, this was a struggle between the "Greek and the Muslim world" that lost the nation's greatest possession, Asia Minor, to the Seljuks.[33] For Ioannes Karagiannopoulos (1923–2000), a professor of Byzantine History at the Aristotle University of Thessaloniki (since 1963) and a founding member of the Centre for Byzantine Studies in Thessaloniki (in 1966), "the defeat of the Byzantines at Manzikert in 1071, despite it being a tragedy, did not matter that much and was not the main cause of all the evils that befell the Greeks of Asia Minor and the empire in general in the aftermath of the battle." We also see the same attempt to celebrate Diogenes as a "martyr-like" figure over his victorious but "petty" [μικρόψυχος] enemy.[34]

Perhaps the most influential academic figure on the Greco-Turkish relations in the second half of the twentieth century was the Greek-American Speros Vryonis, Jr. (1928–2019). He worked and taught in several institutions in the USA (Dumbarton Oaks, Harvard, and UCLA) and briefly in Greece (University of Athens). Vryonis's research interests included the history of the Byzantine Empire, the relations between

30 Amantos, *Σχέσεις Ελλήνων και Τούρκων*, 30–31.

31 Kitromilides, "On the Intellectual Content of Greek Nationalism," 25n1.

32 Μαρία Νυσταζοπούλου-Πελεκίδου, "Οἱ Βυζαντινὲς ἱστορικὲς σπουδὲς στὴν Ελλάδα: Ἀπὸ τὸν Σπυρίδωνα Ζαμπέλιο στὸν Διονύσιο Ζακυθηνό," *Byzantina Symmeikta* 9 (1994): 173–76.

33 Zakythinos, *Βυζαντινή Ιστορια* 324–1071, 492–93.

34 Ioannes Karagiannopoulos, *Ιστορία του Βυζαντινού Κράτους* (Thessaloniki: Vanias, 1976, 1978, 1990), 2:579–81.

the neighbouring ethnic groups of Asia Minor (Armenians, Greeks, and Turks), the relations between the Balkan peoples, and the ideological and political use of history in modern Turkey. In his monumental *The Decline of Medieval Hellenism in Asia Minor and the Process of Islamization from the Eleventh through the Fifteenth Century* (Berkeley: University of California Press, 1971), the title of the voluminous study leaves no doubt about the attitude with regards to the defeat of the Byzantine arms and the decline of the glorious empire and its withdrawal from its natural habitat by the enemy from the East.[35] Finally, Helene Glykatzi-Ahrweiler (b. 1926), one of the greatest Greek Byzantinists and intellectuals of the twentieth century, wrote about the Battle of Manzikert in one of her many studies about the Byzantine Empire, "Byzantium was thus trapped between East and West from the end of the 11th century onwards [the battle and the shocking defeat at Manzikert in 1071 coincided with the fall of Bari, the last Byzantine possession in Italy], the former world power, Byzantium, became mortally wounded in 1204 and breathed its last breath in 1453, sweeping away with the fall of New Rome the beliefs and prophecies of its eternity."[36]

Not a professional historian but a distinguished novelist, Penelope Delta (1874–1941) is widely celebrated for her contributions to the field of children's literature. Her historical novels have been widely read and have influenced popular modern Greek perceptions of national identity and history. I was moved when I read her *In the Heroic Age of Basil II: Emperor of Byzantium* (originally "Τον καιρό του Βουλγαροκτόνου," Athens: 1911) in elementary school, which may have contributed a small part in my decision to focus in the "glorious" period of the Macedonian dynasty. Yet, I was completely unaware of her unfinished novel *Το Γκρέμισμα* [The Tearing].[37] The narrative begins on the night of the Battle of Manzikert, following the pitched battle, a date which according to Delta "is the beginning of the end; the hour of the death of the Byzantine state," where the captured Romanos Diogenes is confounded in a makeshift tent, strictly guarded by a Turkish slave-soldier, defeated, when he could have been the victor, he recalls all that has taken place from the day of his accession to the throne to this moment, when, exhausted, awaits the decision of the victor, the Turkish Sultan Alp Arslan, about his fate.

In 1912, Penelope Delta began to study the era of Alexios Komnenos and write *The Tearing*, and met the Greek Prime Minister Eleftherios Venizelos for the first time in Athens, with Delta's father Emmanuel Benakis being Venizelos's finance minister. She worked on her novel for almost fifteen years, until she finally gave up in 1927 and refused to publish it (it was published only in 1983). This is a critical period for the history of Greece and the Greek–Turkish relations and, going back to her correspondence from 1921 to 1927, a period during which she attempted to complete *The Tearing*, but she abandoned the idea for good; for Delta, the choice of the title is reflected in the "tearing" of the Greeks from Asia Minor in 1922, and the "fall" of Eleftherios Venizelos and

35 Vryonis Jr., *παρακμή του μεσαιωνικού ελληνισμού*, 92–93.

36 Helene Glykatzi-Ahrweiler, *Γιατί το Βυζάντιο* [Why Byzantium] (Athens: Ελληνικά Γράμματα, 2009), 252.

37 The following pages are based on: Vaia Kapnia, "Προσεγγίσεις στο Ιστορικό Μυθιστόρημα της Πηνελόπης Δέλτα" (PhD diss., University of Thessaly, 2009), especially 50–70.

modern Greece is reflected in the fate of Romanos Diogenes and the Byzantine Empire after 1071.

The figure of a charismatic yet tragic leader is a powerful tool in Delta's literary canvas. In each of her works she uses a historical context that was glorious, victorious, and uplifting for the Greek nation, whether he was Basil II "the Bulgar Slayer" (the nickname became widely used by the Greeks during the so-called "Macedonian Question" between 1893 and 1912 that led to the First and Second Balkan Wars of 1912–1913), Romanos Diogenes, Alexios Komnenos, or the "Makedonomachos" Tellos Agras, the hero who for Delta is the basis of Greece's success. Bravery, courage, leadership, foresight, intelligence, and magnanimity were the hallmarks of these historical figures, and Delta recognized them in Eleftherios Venizelos. However, as soon as there is no *pistis* ("loyalty") to the leader, there comes the anticipated fall/decline for the people, just like with Romanos and Hellenism after Manzikert, which is paralleled with the 1920 electoral defeat of the "liberal" Venizelos, his subsequent self-exile to Paris, and the "Great Disaster" that followed the defeat of the Greek army by the Kemalist troops in 1922.

The Reception of the Battle of Manzikert in Modern Turkish Historiography

The birth of the legend surrounding the Battle of Manzikert and the triumph of the Seljuk Sultan Alp Arslan in modern Turkish historiography owes its origins to the turbulent political developments in the Balkans at the turn of the twentieth century. The reference here is to the so-called "Macedonian Question" and the eventual loss of the majority of the Ottoman Balkan provinces of Rumelia to Greece, Serbia, and Bulgaria during the Balkan Wars of 1912–1913 that eventually stimulated a new interest in Anatolia, which was rendered the "homeland of the Turks."[38] It also arguably emerged as a reaction to the utopian "Pan-Turkism" and visions of the mythical Turan, which some regarded as an artificial construct promoted by the "Tatar intellectuals" who settled in Istanbul after 1908.[39]

The historiographic interest in the Anatolian Seljuks that emerged in the 1910s was further advanced in the first two decades of the Republic, as it was understood to be a time when pre-Ottoman themes would become increasingly disassociated from the (Ottoman) dynastic structures of politics for Turkish intellectuals. Hence, we have the first articles on the post-Seljukid and pre-Ottoman Turkoman principalities by Ahmed Tevfik[40] in the *Tarih-i Osmani Encümeni Mecmuasi* in the 1910s. A burgeoning interest in

38 Sina Aksin, "Three Homelands of the Turks," in *Essays in Ottoman-Turkish Political History*, ed. Sina Aksin (Piscataway: Gorgias, 2011), 137–44; Can Erimtan, "Hittites, Ottomans and Turks: Ağaoğlu Ahmed Bey and the Kemalist Construction of Turkish Nationhood in Anatolia," *Anatolian Studies* 58 (2008): 141–71.

39 Doğan Gürpinar, "Anatolia's Eternal Destiny was Sealed: Seljuks of Rum in the Turkish National(ist) Imagination from the Late Ottoman Empire to the Republican Era," *European Journal of Turkish Studies* (online), May 2, 2012, 15; Erimtan, "Hittites, Ottomans and Turks," 156–57.

40 Christoph Herzog, "Ahmed Tevfik Paşa, Okday," in *Encyclopaedia of Islam Online*, http://dx.doi.org/10.1163/1573-3912_ei3_SIM_0318, accessed October 31, 2022.

the local histories of Anatolia, rather than being regarded as subverting Turkish national identity and unity, it was viewed as reinforcing the allegiance of the local populations to the "national narrative," with İsmail Hakki Uzunçarşili publishing several monographs on the cultural heritage of the Seljuks and the post-Seljukid Turkoman principalities in the provincial centres of Anatolia.[41] Moreover, during the early years of the Turkish Republic the name of the polity was changed from *Rum Selçuklulari* [Seljuks of Rum] to *Anadolu Selçuklulari* [Anatolian Seljuks] to further "Turkify" the sovereign territory of Turkey by dismissing the term Rum that had implicit associations with the Byzantine-Greek history of the region.[42] The Kemalist historiography perceived the Anatolian Seljuks as the predecessors and historical "role models" of the Ottoman Empire, thus encouraging the perception that Anatolia was the manifest destiny for the Turks.[43]

For Yahya Kemal (1884–1958), a leading Turkish politician, poet, and author, Turkish history began with Manzikert, with the penetration of Anatolia by the Turks and the subsequent intermarriages thus creating the conditions for the emergence of a new nation; it was Manzikert that made the Turks and not vice versa.[44] Abdurrahman Şeref (1853–1925)[45] mentioned the Byzantine campaigns that culminated at Manzikert as "the war near Manzikert" where the "Roman emperor Roman Diogenes was defeated and taken captive" without any reference to the "opening of Anatolia."[46] Ahmed Cevdet Pasha (1822–1895),[47] an Ottoman scholar, intellectual, bureaucrat, administrator, and historian who was a prominent figure in the Tanzimat reforms of the Ottoman Empire, he also only referred to the defeat of the Byzantine army and the captivity of Romanos Diogenes, but did not place any geopolitical importance on the outcome of the Battle of Manzikert.[48] Necib Asim (1861–1935),[49] an officer of the Ottoman army, scholar, historian, Turkologist, and a MP, he "secularized" the battle and described it as "an immense and definite success" [büyük ve kati bir muvaffakiyete nail olan].[50] Ahmed Refik focused

---

41 Mehmet İpşirli, "Uzunçarşili, Ismail Hakki," in vol. 42 (Tütün–Vehran) of *TDV Encyclopedia of Islam* (in Turkish) (Istanbul: Turkiye Diyanet Foundation, 2012), https://islamansiklopedisi.org.tr/uzuncarsili-ismail-hakki, accessed October 31, 2022.

42 Fuad Köprülü employed the terms Rum Selçuklulari and Anadolu Selçuklulari interchangeably, sometimes referring to the polity simply as Anadolu. Gürpinar, "Anatolia's Eternal Destiny," 11n18.

43 Gürpinar, "Anatolia's Eternal Destiny," 13; Erimtan, "Hittites, Ottomans and Turks," 162–67.

44 Gürpinar, "Anatolia's Eternal Destiny," 17.

45 Abdülkadir Özcan, "Abdurrahman Şeref," in *TDV Encyclopedia of Islam*, https://islamansiklopedisi.org.tr/abdurrahman-seref, accessed October 31, 2022.

46 Abdurrahman Şeref, *Fezleke-i Tarih-i Devlet-i Osmaniye* (Istanbul: Karabet Matbaasi, 1315/1897), 1:47; Gürpinar, "Anatolia's Eternal Destiny," 16.

47 Osman G Özgüdenli, "Jevdet Pasha," in *Encyclopædia Iranica Online*, https://iranicaonline.org/articles/jevdet-pasha, accessed October 31, 2022.

48 Ahmed Jevdet Pasha, *Tarih-i Cevdet* (Istanbul: Matbaa-I Osmaniye, 1309/1891–92), 1:26; Gürpinar, "Anatolia's Eternal Destiny," 16.

49 Abdullah Uçman, "Necip Âsim Yaziksiz," in *TDV Encyclopedia of Islam*, https://islamansiklopedisi.org.tr/necip-asim-yaziksiz, accessed October 31, 2022.

50 Necib Asim, Türk Tarihi (Istanbul 1316/1898), 271; Gürpinar, "Anatolia's Eternal Destiny," 16.

a significant part of his *Tarih-i Umumi* [General History] to the Seljuks of Rum, yet [51] he did not defer Manzikert with any substantial geopolitical outcome, rather emphasizing the role of Süleyman Ibn Kutulmuş in the "Turkification" of Anatolia in the 1070s.[52]

Only in 1918 did Fuad Köprülü (1890–1966) establish a connection between Manzikert and the Turkish penetration and settlement in Anatolia.[53] As a young and highly educated Turkish nationalist, he committed himself to proving the genuine Turkishness of the Ottomans, and the Seljuks of Rum played an instrumental role in this as a world that had not become corrupted by Arab and Persian influences. In among the first works he produced, he wrote that "Completely defeating the Roman emperor Romanos Diogenes in the Battle of Manzikert, [Alp Arslan...established absolute Turkish rule in Anatolia."[54] Four years later, Köprülü referred to the battle as "the definite victory of Manzikert" in "Turkifying" Anatolia, while in 1926 he wrote in his *History of Turkish Literature* that "the victory in Manzikert opened all the paths to Anatolia."[55]

Mükrimin Halil Yinanç's *Türkiye Tarihi, Selçuklular Devri I, Anadolu'nun Fethi* [History of Turkey, Seljuks Period I, the Conquest of Anatolia], published in Istanbul in 1944, highlighted the Battle of Manzikert not only as the turning point in Turkic history,[56] but also "a turning point in world history."[57] Yinanç was one of the proponents of the Anatolian ideology in the 1920s and emerged as a pioneer in the scholarly study of the Seljuks of Rum in Turkey. For Mehmet Feridun Dirimtekin (1894–1976),[58] a staff army officer and an ardent Kemalist, Manzikert was the "Turkish Battle of Cannae" and a "turning point in world history."[59]

The Battle of Manzikert gained much greater prominence in Kemalist historiography in the 1960s and 1970s with the emergence of the composite of Islamic and national-

---

51 Zeki Arikan, "Altinay, Ahmed Refik," in *Encyclopaedia of Islam Online*, http://dx.doi.org/10.1163/1573-3912_ei3_COM_24194, accessed October 31, 2022; Abdülkadir Özcan, "Ahmet Refik Altinay" in *TDV Encyclopedia of Islam*, https://islamansiklopedisi.org.tr/ahmed-refik-altinay, accessed October 31, 2022.

52 Ahmed Refik, *Tarih-i Umumi* (Istanbul: Kitabhane-i İslam ve Askeri, 1328/1910), 6:286; Gürpinar, "Anatolia's Eternal Destiny," 17.

53 Ömer Faruk Akün, "Mehmed Fuad Köprülü", in *TDV Encyclopedia of Islam*, https://islamansiklopedisi.org.tr/mehmed-fuad-koprulu, accessed October 31, 2022.

54 Mehmed Fuad Köprülü, Türk Edebiyatinda İlk Mutasavviflar (Istanbul: Matba-I Amire, 1918), 208; Gürpinar, "Anatolia's Eternal Destiny," 17.

55 Mehmed Fuad Köprülü, "Anadolu'da İslamiyet: Türk İstilasindan Sonra Anadolu Tarih-i Dinisine Bir Nazar ve Bu Tarihin Menbalari," *Darülfünun Edebiyat Fakültesi Mecmuasi* 2 (1338–39): 282; Gürpinar, "Anatolia's Eternal Destiny," 17.

56 Fahameddin Başar, "Yinanç, Mükrimin Halil," in *TDV Encyclopedia of Islam*, https://islamansiklopedisi.org.tr/yinanc-mukrimin-halil, accessed October 31, 2022.

57 Mükrimin Halil Yinanç, Türkiye Tarihi, Selçuklular Devri I, *Anadolu'nun Fethi* (Istanbul, 1944), 78; Gürpinar, "Anatolia's Eternal Destiny," 17.

58 https://ataturkansiklopedisi.gov.tr/bilgi/mehmet-feridun-dirimtekin-1894-1976/, accessed October 31, 2022.

59 Mehmet Feridun Dirimtekin, *Malazgirt Meydan Muharebesi, 26 Ağustos 1071* (Istanbul: Ahmet Halit Kitabevi, 1936/1943), 1; Gürpinar, "Anatolia's Eternal Destiny," 17.

ist ideas known as the "Turco-Islamic synthesis" that sprang up as the nine-hundredth anniversary of the battle drew nearer. During the early 1970s, several conservative intellectuals reconciled Turkish nationalism with Islam, rejecting the strictly secular interpretations of Turkish nationalism and arguing that Islam and Turkish identity could complement each other.[60] A prominent figure in this process was İbrahim Kafesoğlu (1914–1984) one of the foremost Turkish historians, a Turkologist, and a prominent academic.[61] One of the ideologists of the Turkish-Islamic synthesis and a well-known historian of the Seljuks of Rum, Kafesoğlu emerged as the chief ideologue of The Institute for Research on Turkish Culture founded in 1961. Kafesoğlu celebrated Alp Arslan as the leader who "exterminated the currents detrimental to Sunni Islam," the one who "maintained political unity" and who "guided the Turks to their promised land, Anatolia, on an Islamic and historical mission."[62]

A contemporary of Kafesoğlu who was also influenced by Fuad Köprülü's article "Anadolu Selçuklulari Tarihi'nin Yerli Kaynaklari" [The Indigenous Sources of the History of the Anatolian Seljuks] and devoted most of his studies to the Anatolian Seljuks was Osman Turan (1914–1978),[63] a brilliant historian who believed that the foundations of Turkish national culture were laid in Central Asia and were transformed with new elements by Islam, and that this culture was uninterrupted. For Turan, "The historical consequences of the Battle of Malazgirt were enormous and constituted a turning point, and it is impossible to show its great impact here."[64] Indeed, for Turan and his esteemed colleague Ali Sevim, Anatolia is identified as a "Turkish homeland" (Anadolu'nun bir Türk yurdu),[65] while for the latter "the Malazgirt Victory was one of the turning points of Turkish and world history and an important event that constitutes one of the most important events of the [Byzantine–Turkish] war," to add further down that the early Turkish settlements in Anatolia in the 1070s "played an important historical role in the transformation of Anatolia into a Turkish homeland."[66]

Speaking at an event held on August 26, 2017 to commemorate the Battle of Manzikert, the President of the Turkish Republic, Recep Tayyip Erdoğan, said:

> With the victory he won in Malazgirt, Sultan Alp Arslan took a step that made Anatolia our eternal homeland. We are proud of our ancestors who walked with glory, honour, and victory until the middle of Europe after entering Ana-

60 Gürpinar, "Anatolia's Eternal Destiny," 18; Hillenbrand, *Turkish Myth*, 210–11.

61 Abdülkadir Donuk, "Kafesoğlu, Halil İbrahim," in *TDV Encyclopedia of Islam*, https://islamansiklopedisi.org.tr/kafesoglu-halil-ibrahim, accessed October 31, 2022.

62 İbrahim Kafesoğlu, Harezmşahlar Devleti Tarihi (Ankara, 1956/1984); Gürpinar, "Anatolia's Eternal Destiny," 19.

63 Salim Koca, "Turan, Osman," in *TDV Encyclopedia of Islam*, https://islamansiklopedisi.org.tr/turan-osman, accessed October 31, 2022.

64 Turan, *Selçuklular Tarihi*, 187 and 299.

65 Sevim, *Anadolu'nun Fethi* Selçuklular Dönemi, 13.

66 Sevim, *Anadolu'nun Fethi* Selçuklular Dönemi, 75.

> tolia from Malazgirt with the red flag in one hand and the green sanjak in the other. This victory gained our people a new homeland and future.

Later in his speech, President Erdoğan added that "nearly two centuries later [after Manzikert], the Ottoman State, one of the largest states that history has ever known, began to take root in Anatolia and gathered four continents under its shadow," thus linking the Seljuk penetration of Anatolia after Manzikert to the Ottoman establishment in the thirteenth and fourteenth centuries, an era that culminated into "a new age once the Ottoman flag started to wave on the walls of Byzantium."[67] Finally, it would be interesting to note that the Çamlica Mosque in Istanbul's Asian side, the largest mosque in Turkey and one of several megaprojects delivered by the ruling AK Party in 2019, has six minarets, four of which have three balconies that stand 107.1 metres high, clearly as a reference to the Battle of Manzikert, while the conference hall of the mosque is able to seat up to 1,071 people, yet another reference to the date of the battle.[68]

---

67 https://www.tccb.gov.tr/en/news/542/83274/victory-of-malazgirt-gained-our-people-a-new-homeland-and-future, accessed October 31, 2022.

68 *Daily Sabah*, published October 8, 2019; accessed October 31, 2022. https://www.dailysabah.com/arts-culture/2019/10/08/camlica-mosque-the-most-modern-complex-of-its-kind-in-turkey.

SELECT BIBLIOGRAPHY

Primary Sources

al-ʿAdīm, Ibn. In Carole Hillenbrand, *Turkish Myth and Muslim Symbol: The Battle of Manzikert,* 74–77. Edinburgh: Edinburgh University Press, 2007.

al-Athīr, Ibn. In Carole Hillenbrand, *Turkish Myth and Muslim Symbol: The Battle of Manzikert,* 64–66. Edinburgh: Edinburgh University Press, 2007.

al-Bundārī. In Carole Hillenbrand, *Turkish Myth and Muslim Symbol: The Battle of Manzikert,* 58–63. Edinburgh: Edinburgh University Press, 2007.

al-Din, Rashid. In Carole Hillenbrand, *Turkish Myth and Muslim Symbol: The Battle of Manzikert,* 64–66. Edinburgh: Edinburgh University Press, 2007.90–95.

al-Fāriqī, Ibn al-Azraq. In Carole Hillenbrand, *Turkish Myth and Muslim Symbol: The Battle of Manzikert,* 33–34. Edinburgh: Edinburgh University Press, 2007.

al-Ḥusaynī. In Carole Hillenbrand, *Turkish Myth and Muslim Symbol: The Battle of Manzikert,* 52–57. Edinburgh: Edinburgh University Press, 2007.

al-Jawzī, Ibn. In Carole Hillenbrand, *Turkish Myth and Muslim Symbol: The Battle of Manzikert,* 37–43. Edinburgh: Edinburgh University Press, 2007.

al-Jawzī, Sibṭ Ibn. In Carole Hillenbrand, *Turkish Myth and Muslim Symbol: The Battle of Manzikert,* 67–73. Edinburgh: Edinburgh University Press, 2007.

al-Qalānisī, Ibn. In Carole Hillenbrand, *Turkish Myth and Muslim Symbol: The Battle of Manzikert,* 30–32. Edinburgh: Edinburgh University Press, 2007.

al-Qalānisī, Ibn. *The Damascus Chronicle* of the Crusades. Partial translation by H. A. R. Gibb. Mineola: Dover, 2002.

First published 1932 Luzac (London).

al-Ṭurṭūshī. In Carole Hillenbrand, *Turkish Myth and Muslim Symbol: The Battle of Manzikert,* 27–29. Edinburgh: Edinburgh University Press, 2007.

The Annals of the Saljuq Turks: Selections from al-Kamil fi'l-Ta'rikh of Ibn al-Athir. Translated by D. S. Richards. London: Routledge, 2002.

"The Anonymous Byzantine Treatise on Strategy." In *Three Byzantine Military Treatises.* Edited and translated by George T. Dennis. Washington, DC: Dumbarton Oaks, 2008.

Aristakes Lastivertc'i's History. Edited by Robert Bedrosian. New York: Sources of the Armenian Tradition, 1985.

Armenia and the Crusades: Tenth to Twelfth Centuries the Chronicle of Matthew of Edessa. Edited by Ara Edmond Dostourian. Lanham: University of America Press, 1993.

Attaleiates, Michael. *Historia.* Edited and translated by Inmaculada Pérez-Martín. Madrid: Consejo Superior de Investigaciones Científicas, 2002.

——. *The History.* Edited and translated by Anthony Kaldellis and Dimitris Krallis. Cambridge: Harvard University Press, 2012.

Βρυέννιος, Νικηφόρος. *Ὕλη Ιστορίας* [Material of History]. Edited and translated by Demetres Tsougkarakes and Despoina Tsouklidou. Athens: Kanake, 1996.

Byzantium in the Time of Troubles: The Continuation of the Chronicle of John Skylitzes (1057–1079). Edited and translated by Eric McGeer. Leiden: Brill, 2020.

The Chronicle of Ibn al-Athīr for the Crusading Period from al-Kāmil fī'l-ta'rīkh. Part 1. Translated by D. S. Richards. New York: Routledge, 2016.

The Chronicle of Ibn al-Athīr for the Crusading Period from al-Kāmil fī'l-ta'rīkh. Part 2. Translated by D. S. Richards. New York: Routledge, 2016.

The Chronicle of Michael the Great, Patriarch of the Syrians. Edited by Robert Bedrosian. Long Branch: Sources of the Armenian Tradition, 2013.

Constantine Porphyrogenitus: Three Treatises on Imperial Military Expeditions. Edited and translated by John F. Haldon. Vienna: Österreichischen Akademie der Wissenschaften, 1990.

The History of Leo the Deacon: Byzantine Military Expansion in the Tenth Century. Translated by D. Sullivan and A-M. Talbot. Washington DC: Dumbarton Oaks, 2005.

John Skylitzes: A Synopsis of Byzantine History, 811–1057. Edited and translated by John Wortley. Cambridge: Cambridge University Press, 2010.

Maurice's Strategikon: Handbook of Byzantine Military Strategy. Edited and translated by George T. Dennis. Philadelphia: University of Pennsylvania Press, 1984.

Mīrkhwānd. In Carole Hillenbrand, *Turkish Myth and Muslim Symbol: The Battle of Manzikert*, 98–104. Edinburgh: Edinburgh University Press, 2007.

Smbat Sparapet's Chronicle. Edited and translated by Robert Bedrosian. Long Branch: Sources of the Armenian Tradition, 2005.

The Taktika of Leo VI. Edited and translated by George T. Dennis. Washington, DC: Dumbarton Oaks, 2010.

Vegetius: Epitome of Military Science. Translated with notes and introduction by N. P. Milner. Liverpool: Liverpool University Press, 2001.

Ζωναράς, Ιωάννης. *Επιτομή Ιστοριών* [Epitome of Histories]. Translated by Iordanes Gregoriades. 3 vols. Athens: Kanake, 1995.

Selected Secondary Sources

Angold, Michael. "Belle Époque or Crisis? (1025–1118)." In *The Cambridge History of the Byzantine Empire, c. 500–1492*. Edited by Jonathan Shepard, 583–626. Cambridge: Cambridge University Press, 2008.

——. *The Byzantine Empire 1025–1204: A Political History*. London: Longman, 1985 [1997].

Beihammer Alexander Daniel. *Byzantium and the Emergence of Muslim-Turkish Anatolia, ca. 1040–1130*. London: Routledge, 2017.

Bosworth, Clifford E. *The History of the Seljuq State: A Translation with Commentary of the Akhbār Aldawla al-Saljūqiyya*. London: Routledge, 2011.

Cahen, Claude. "Le Malik-Nameh et l'histoire des origines Seljukides." *Oriens* 2 (1949): 31–65.

——. *Pre-Ottoman Turkey: A General Survey of the Material and Spiritual Culture and History c. 1071–1330*. New York: Taplinger, 1968.

Cheynet, Jean Claude. "Mantzikert: Un désastre militaire?" *Byzantion* 50 (1980): 410–38.

Dostourian, Ara Edmond. "The Chronicle of Matthew of Edessa (Matteos Urhayetsi)." In *Armenian Tigranakert and Diyarbakir/Edessa*. Edited by R. G. Hovannisian, 155–64. Costa Mesa: Mazda, 2006.

Gautier, Paul. *Nicephore Bryennios Histoire*. Brussels: Byzantion, 1975.

Gregoriades, Iordanes. *Linguistic and Literary Studies in the Epitome Historion of John Zonaras*. Thessaloniki: Centre for Byzantine Research, 1998.

Haldon, John F. "Approaches to an Alternative Military History of the Period ca. 1025–1071." In *Η αυτοκρατορία σε κρίση: Το Βυζάντιο τον 11ο αι. (1025–1081)* [The Empire in Crisis(?): Byzantium in the Eleventh Century (1025–1081)]. Edited by Vassiliki B. Vlysidou, 45–74. Athens: National Hellenic Research Foundation, 2003.

——. *The Byzantine Wars*. Stroud: Tempus, 2001.

——. "La logistique de Mantzikert." In *Guerre et société au Moyen âge: Byzance-Occident (VIIIe–XIIIe siècle)*. Edited by Dominique Barthélemy and Jean-Claude Cheynet, 11–26. Paris: Association des amis du Centre d'histoire et civilisation de Byzance, 2010.

——. *Warfare, State and Society in the Byzantine World, 565–1204*. London: University College of London Press, 1999.

——, Vince Gaffney, Georgios Theodoropoulos, and Phil Murgatroyd. "Marching Across Anatolia: Medieval Logistics and Modeling the Mantzikert Campaign." *Dumbarton Oaks Papers* 65/66 (2011–2012): 209–35.

Harari, Yuval Noah. "The Concept of 'Decisive Battles' in World History." *Journal of World History* 18 (2007): 251–66.

Hillenbrand, Carole. *Turkish Myth and Muslim Symbol: The Battle of Manzikert*. Edinburgh: Edinburgh University Press, 2007.

Howard-Johnston, James. "Anna Komnene and the *Alexiad*." In *Alexios I Komnenos*. Edited by Margaret Mullett and Dion Smythe, 260–302. Belfast: Priory, 1996.

Hunger, Herbert. *Βυζαντινή λογοτεχνία: η λόγια κοσμική γραμματεία των Βυζαντινών* [Byzantine Literature: The Secular Literature of the Byzantines]. Translated by Taxiarchis Kolias, Katerina Syneli, G. X. Makris, and Ioannis Bassis. Athens: MIET, 1997.

Izutsu, Toshihiko. *Ethico-Religious Concepts in the Qur'an*. Montreal: McGill-Queen's University Press, 2002.

Jeffreys, Elizabeth M. "Nikephoros Bryennios Reconsidered." In *Η αυτοκρατορία σε κρίση: Το Βυζάντιο τον 11ο αι. (1025–1081)* [The Empire in Crisis(?): Byzantium in the Eleventh Century (1025–1081)]. Edited by Vassiliki B. Vlysidou, 201–14. Athens: National Hellenic Research Foundation, 2003.

Kagan, Kimberly. *The Eye of Command*. Ann Arbor: University of Michigan Press, 2006.

Kaldellis, Anthony. *Ethnography After Antiquity: Foreign Lands and Peoples in Byzantine Literature*. Philadelphia: University of Pennsylvania Press, 2013.

Karpozilos, Apostolos. *Βυζαντινοί Ιστορικοί και Χρονογράφοι (11ος–12ος αι.)* [Byzantine Historians and Chronographers (Eleventh–Twelfth Centuries)]. Athens: Kanake, 2009.

Keegan, John. *The Face of Battle: A Study of Agincourt, Waterloo and the Somme*. London: Pimlico, 2004.

Kiapidou, Eirene-Sofia, *Η Σύνοψη Ιστοριών του Ιωάννη Σκυλίτζη και οι πηγές της, (811–1057) Συμβολή στην Βυζαντινή Ιστοριογραφία κατά τον ΙΑ'αιώνα* [The Synopsis of Histories of John Skylitzis and its Sources, (811–1057): Contribution to Byzantine Historiography in the 1st Century]. Athens: Kanake, 2010.

Kühn, Hans Joachim. *Die byzantinische Armee im 10. und 11: Jahrhundert: Studien zur Organisation der Tagmata*. Vienna: Fassbaender, 1991.

Leveniotes, Georgios A. "Η πολιτική κατάρρευση του Βυζαντίου στην Ανατολή: το ανατολικό σύνορο και η κεντρική Μικρά Ασία κατά το β' ήμισυ του 11ου αι" [The Political Collapse of Byzantium in the East: The Eastern Frontier and Central Asia Minor in the Second Half of the Eleventh Century]. PhD diss., Aristotle University of Thessalonike, 2007.

——. *Το στασιαστικό κίνημα του Νορμανδού Ουρσελίου (Ursel de Bailleul) στη Μικρά Ασία (1073–1076)* [The Rebellious Movement of Ursel de Bailleul in Asia Minor (1073–1076)]. Thessaloniki: Vanias, 2004.

Macrides, Ruth. "The Historian in the History." In *Philhellen: Studies in Honour of Robert Browning*. Edited by Costas N. Constantinides, Nikolaos M. Panagiotakes, Elizabeth Jeffreys, and Athanasios D. Angelou, 205–24. Venice: Istituto Ellenico di Studi Bizantini e Postbizantini di Venezia, 1996.

Marincola, John. *Authority and Tradition in Ancient Historiography*. Cambridge: Cambridge University Press, 1997.

Matheou, Nikolas. "Situating the History Attributed to Aristakēs Lastiverc'i, 1000–1072: The Empire of New Rome and Caucasia in the Eleventh Century." PhD diss., University of Oxford, 2018.

McGrath, Stamatina. "Warfare as Literary Narrative." In *A Companion to the Byzantine Culture of War, ca. 300–1204*. Edited by Yannis Stouraitis, 160–95. Leiden: Brill, 2018.

Miotto, Marco. *Ο Ανταγωνισμός Βυζαντίου Και Χαλιφάτου Των Φατιμιδών Στην Εγγύς Ανατολή και η δράση των Ιταλικών πόλεων στην περιοχή κατά το 10ο και τον 11ο αιώνα* [The Competition between Byzantium and the Fatimid Caliphate in the Near East and the Activity of Italian Cities in the Region in the Tenth and Eleventh Centuries]. Thessaloniki: Kentro Byzantinon Ereunon, 2008.

Neville, Leonora. *Guide to Byzantine Historical Writing*. Cambridge: Cambridge University Press, 2018.

——. *Heroes and Romans in Twelfth-Century Byzantium: The "Material for History" of Nikephoros Bryennios*. Cambridge: Cambridge University Press, 2012.

——. "Why Did the Byzantines Write History?" In *Proceedings of the 23rd International Congress of Byzantine Studies Belgrade, 22–27 August 2016*. Edited by Smilja Marjanović-Dušanić, 265–76. Belgrade: Serbian National Committee of AIEB, 2016.

Nicolle, David. *Manzikert 1071: The Breaking of Byzantium*. Oxford: Osprey, 2013.

Peacock, A. C. S. *Early Seljuq History, A New Interpretation*. London: Routledge, 2013.

Ramsay, William M. *The Historical Geography of Asia Minor*. 4 vols. London: John Murray, 1890.

Robinson, Chase F. *Islamic Historiography*. Oxford: Oxford University Press, 2003.

Sevim, Ali. *Anadolu'nun Fethi Selçuklular Dönemi* [Conquest of Anatolia Seljuk Period]. Ankara: Türk Tarih Kurumu basimevi, 1988.

Sinclair, Kyle James. "War Writing in Middle Byzantine Historiography: Sources, Influences and Trends." PhD diss., University of Birmingham, 2012.

Sinclair, T. A. *Eastern Turkey: An Architectural and Archaeological Survey*. 4 vols. London: Pindar, 1987.

Tara L. Andrews. *Matt'eos Urhayec'i and his Chronicle: History as Apocalypse in a Crossroads of Cultures*. Leiden: Brill, 2017.

Theotokis, Georgios. *Byzantine Military Tactics in Syria and Mesopotamia in the Tenth Century: A Comparative Study*. Edinburgh: Edinburgh University Press, 2018.

Treadgold, Warren. *Byzantium and its Army, 284–1081*. Stanford: Stanford University Press: 1995.

——. *The Middle Byzantine Historians*. New York: Palgrave Macmillan, 2013.

Tsougkarakes, Demetrios. "Η αυτοκρατορία σε κρίση και η οπτική των σύγχρονων: μια ανάγνωση των μαρτυριών" [The Empire in Crisis and the Perspective of Contemporaries: A Reading of the Testimonies]. In *Η αυτοκρατορία σε κρίση: Το Βυζάντιο τον 11ο αι. (1025–1081)* [The Empire in Crisis(?): Byzantium in the Eleventh Century (1025–1081)]. Edited by Vassiliki B. Vlysidou, 275–90. Athens: National Hellenic Research Foundation, 2003.

Turan, Osman, *Selçuklular Tarihi ve Türk-İslâm Medeniyeti* [History of Seljuks and Turkish-Islamic Civilization]. Istanbul: Ötüken Neşriyat, 2008.

Vratimos, Antonios. "Joseph Tarchaneiotes and the Battle of Mantzikert (AD 1071)." *Al-Masāq* 32 (2020): 156–68.

——. "The Two Expeditions of the Byzantine Emperor Romanos IV Diogenes in 1068 and 1069." Ph.D diss., Cardiff University, 2005.

——. "Was Michael Attaleiates Present at the Battle of Mantzikert?" *Byzantinische Zeitschrift* 105 (2012): 829–40.

Vryonis Jr., Speros. "The Eleventh Century: Was There a Crisis in the Empire? The Decline of Quality and Quantity in the Byzantine Armed Forces." In *Η αυτοκρατορία σε κρίση: Το Βυζάντιο τον 11ο αι. (1025–1081)* [The Empire in Crisis(?): Byzantium in the Eleventh Century (1025–1081)]. Edited by Vassiliki B. Vlysidou, 17–43. Athens: National Hellenic Research Foundation, 2003.

——. *Η παρακμή του μεσαιωνικού ελληνισμού στη Μικρά Ασία και η διαδικασία του εξισλαμισμού (11ος–15ος αιώνας)* [The Decline of Medieval Hellenism in Asia Minor and the Process of Islamization (Eleventh–Fifteenth Centuries)]. Athens: MIET, 2008.

——. "A Personal History of the History of the Battle of Mantzikert." In *Byzantine Asia Minor (6th–12th cent.)*. Edited by Stelios Lampakis, 225–44. Athens: National Hellenic Research Foundation, 1998.

Whately, Conor. *Battles and Generals: Combat, Culture, and Didacticism in Procopius' Wars*. Leiden: Brill, 2016.

Yinanç, Mükrimin Halil. *Türkiye Tarihi Selçuklular Devri* [History of Turkey Seljuk Period]. Ankara: Türk Tarih Kurumu, 2020.

INDEX

Printed and bound by CPI Group (UK) Ltd, Croydon, CR0 4YY

15/09/2026

14962387-0001